PLAIN BEARINGS – ENERGY EFFICIENCY AND DESIGN

Tribology Group Committee

Seminar Advisers

PLAIN BEARINGS – ENERGY EFFICIENCY AND DESIGN

Papers presented at a Seminar organized by the Institution
of Mechanical Engineers, and held at the Institution of
Mechanical Engineers on 10 November 1992.

Published by
Mechanical Engineering Publications Limited for
The Institution of Mechanical Engineers
LONDON

Printed by Waveney Print Services Ltd, Beccles, Suffolk

CONTENTS

A finite element model for profile bore bearings and its verification

D T GETHIN, BSc, PhD, MIEE, MIMechE
Department of Mechanical Engineering, University College of Swansea, UK

SUMMARY

The paper describes a comprehensive finite element model to predict the performance of lemon bore and three lobe bearing geometries. The model is compared with published theoretical results for isothermal films and experimental data for thermal analysis. In each case, good agreement has been achieved. Agreement with lemon bore experimental data was obtained when thermal distortion was approximated and heat transfer across the groove was included. Agreement for three lobe bearing behaviour was found to be best when heat transfer over the groove was included, however, this mechanism requires further formal analysis. It was found that loading direction influences bearing thermal behaviour significantly for both geometries.

NOMENCLATURE

C_b base circle clearance (m)

C_f lubricant specific heat capacity (J/kgK)

C_1 lobe clearance (m)

D journal diameter (m)

H power loss (watts)

$\overline{H}$ dimensionless power loss $\left[= \dfrac{C_1 H}{\pi^3 N^2 LD^3} \right]$

$\overline{HI}$ dimensionless power loss $\left[= \dfrac{H}{\mu N^2 LD^2} \left(\dfrac{C_b}{R} \right) \right]$

L bearing length (m)

N rotational speed (rev/s)

P load capacity (N)

$\overline{PI}$ dimension less load $\left[= \dfrac{P}{LD} \dfrac{1}{\mu N} \left(\dfrac{C_b}{R} \right)^2 \right]$

Q flow rate (m^3/s)

Q_{co} flow carried over from the upstream lobe (m^3/s)

Qf infeed flow (m^3/s)

Qin flow entering the lobe (m^3/s)

$\overline{Q}$ dimensionless flow $\left[= \dfrac{2Q}{\pi NDLC_1} \right]$

$\overline{QI}$ dimensionless flow $\left[= \dfrac{2Q}{2C_b LDN} \right]$

R journal radius (m)

R_b base circle radius (m)

R' journal radius corrected for thermal expansion (m)

S Sommerfeld number

T temperature (°C)

Tco temperature of oil carried over from the upstream lobe (°C).

Tf lubricant feed temperature (°C)

Tin temperature of lubricant into the lobe (°C)

T_j journal surface temperature (°C)

To journal temperature on assembly (°C)

T_1 lobe downstream temperature (at the bearing bush outer surface) (°C)

Us sliding velocity (m/s)

a,b constants in Walther equation

e   eccentricity (m)

e_1 lobe eccentricity (m)

h   local film thickness (m)

hs   film thickness at section "s" (m)

ho   film thickness at lobe entry (m)

k_f lubricant thermal conductivity (W/m$\,^\circ$C)

k_s bush thermal conductivity (W/m$\,^\circ$C)

l_s film length at section "s" (m)

m preset $\left(= 1 - \dfrac{C_b}{C_1}\right)$

p   pressure (N/m^2)

u,v velocity components (m/s)

α coefficient of thermal expansion (per $^\circ$C)

δ journal thermal growth (m)

ε eccentricity ratio ($= e/C_b$)

ε_1 lobe eccentricity ratio ($= e_1/C_1$)

μ lubricant dynamic viscosity (kg/ms)

ν lubricant kinematic viscosity (m^2/s)

ρ lubricant density (kg/m^3)

ϕ tilt angle (degrees)

θ loading angle (degrees)

ψ attitude angle

ω rotational speed (rev/min)

1.   INTRODUCTION

Profile bore bearings are used extensively when machinery runs at high rotational speed. They are selected since they can be designed to assure stable shaft operation [1] and this is enabled by their accurate manufacture using computer based technology [2, 3]. The work presented in [1] discusses the stability of the various bearing types under circumstances of different loading direction and highlights the benefit to be derived from tilted three lobe types. Results were also presented for lemon bore geometries which have the benefit of manufacturing simplicity (see Figure 1). With regard to manufacturing, the work discussed in [3] illustrates the advantage of using computer based technology to maintain bearing bore geometric and dimensional tolerance. These latter factors have considerable implications for the speed threshold for the onset of unstable operation. Due to the respective applicability and manufacturing simplicity of three lobe and lemon bore bearings, these will be considered further in this paper.

A number of investigations into the behaviour of profile bore bearings have been completed and which encompass both experimental and theoretical work. Experimental studies have been completed to investigate the stability of profile bore bearings as exemplified by the work described in [4]. In this study, a three rotor system was examined and three bearing types (cylindrical, pressure-dam and three lobe) were tested under circumstances of different loading direction. This work confirmed the superior stability of the three lobe geometry and its marked sensitivity with regard to loading direction. The measured stability was also compared with prediction which was found to underestimate it by between 16% and 34% due to the exclusion of thermal effects which was demonstrated in a discussion on [4] which incorporated a simple heat balance. This confirms the need for a more rigorous treatment of thermal aspects to enable a more accurate prediction of stability under conditions of high speed operation.

From the viewpoint of detail thermal measurements, work pertinent to profile bore geometries has been presented in [5-7]. The work in [5] and [6] has been directed towards large turbo alternator bearings (typical journal diameter = 500mm, L/D = 1.0) and some of the types investigated were lemon bore in form. These studies provide information on the thermal behaviour of the bearing with detail measurements including pressure and film thickness variation (which combines journal location and bearing distortion) presented in [6]. Both studies illustrate the more severe thermal effects associated with superlaminar and turbulent flow and the work reported in [6] confirms the need for rigorous thermohydrodynamic analysis.

Work on a three lobe bearing of smaller scale (journal diameter = 75mm, L/D = 0.53) has been reported in [7]. This includes both symmetric and tilted geometries, combined with a side rail for the tilted geometry. Detail thermal information was recorded along with global performance trends. Again this confirmed the generation of a severe thermal environment at high sliding speed and its dependence on loading direction. The global trends showed that when compared with the symmetric profile, tilt increased flow requirements with a consequent reduction in maximum film temperature. However, the inclusion of the side rail reduced the flow requirement, but with a small increase in power loss due to the thinner film at the bearing edge. Thus, the work described in [5-7] supports the need for a comprehensive model to describe the thermal behaviour of the bearing under the more demanding operating conditions.

A number of comprehensive thermohydrodynamic models of bearing operation have been reported in the literature, see for example [8-10]. The work reported in [8] was one of the first to include a turbulence model with the rigorous prediction of the thermal field in an essentially cylindrical geometry. The model presented in [9] is also comprehensive and accounts for turbulent transport and these were incorporated into a finite difference calculation scheme. Detail and global predicted results were compared with data presented in [6] with good agreement for the film thickness profile measured in the experimental study and which confirms the applicability of the hydrodynamic analysis.

In [10], the thermal analysis was based on a laminar film model, but incorporated a rigorous treatment of the thermal aspects and employed a finite element approach to the solution of the equations. Comparison of detail and global data for a variety of loading conditions was presented and good overall agreement was demonstrated. However, this analysis assumed that the film always occupied the full bearing width which may not be appropriate under all loading directions, particularly for the lemon bore geometries. Such film incompleteness has a significant effect on sideflow as discussed in connection with cylindrical geometries [11].

Since thermal effects are important in journal bearing operation, these will be considered further in this paper and will be addressed towards lemon bore and three lobe bearing geometries. The models developed will be verified by comparing them against currently available data for both isothermal and thermal conditions.

2. MATHEMATICAL BASIS AND NUMERICAL MODEL

In this paper, both isothermal and thermal analyses will be presented, however the explanation to be presented below will focus on a thermal model since the isothermal analysis is a simplified subset of this.

The bearing model may be simplified by considering the thermal behaviour on the centreplane only and applying this over the bearing width since the axial temperature variation over the bearing width is usually small [7]. Figure 2a illustrates a bearing lobe which clearly exceeds an included angle of $180°$, but which shows the features which are associated with both divergent and convergent films. Notably where convergence occurs there will be spreading of the lubricant film until it fills the full bearing width whereas for divergence some of the film width may become incomplete dependent on the oil flow through the thinnest section. These mechanisms need to be included in the numerical model appropriate to the film thickness changes in each lobe. This is an idealised representation and

neglects the generation of cavitation streamers which occur in actual bearing operation [12].

To account for temperature (and hence viscosity) variation through the film, the most economic solution for the hydrodynamic action is based on the generalised pressure equation [13] which may be written as:

$$\frac{\partial}{\partial x}\left[G \frac{\partial p}{\partial x} \right] + \frac{\partial}{\partial z}\left[G\frac{\partial p}{\partial z} \right] =$$

$$Us \frac{dh}{dx} - Us \frac{dF}{dx} \qquad \ldots (1)$$

$$G = \int_0^h \frac{y}{\mu} (y - F)\, dy; \quad F = \frac{F_1}{F_0}; \quad F_0 =$$

$$\int_0^h \frac{1}{\mu}\, dy; \quad F_1 = \int_0^h \frac{y}{\mu}\, dy$$

In the sliding direction, the velocity variation over the film thickness is given by:

$$u(Y) = \left[\frac{Us}{F_0} - \frac{F_1}{F_0} \frac{\partial p}{\partial x} \right] \int_0^Y \frac{1}{\mu}\, dy +$$

$$\frac{\partial p}{\partial x} \int_0^Y \frac{1}{\mu}\, dy \qquad \ldots (2)$$

while transverse to the film plane, the velocity is given by:

$$v(Y) = \left(\frac{Y}{h}\right)^2 Us\, Sin \frac{dh}{dx} \qquad \ldots (3)$$

For the thermal analysis, the energy equation on the bearing centreplane can be written as

$$\rho C_f \left[u \frac{\partial T}{\partial x} + v \frac{\partial T}{\partial y} \right] = k_f \left[\frac{\partial^2 T}{\partial x^2} + \frac{\partial^2 T}{\partial y^2} \right] +$$

$$\mu \left[\frac{\partial u}{\partial y} \right]^2 \qquad \ldots (4)$$

and in the bearing bush, this simplifies to

$$k_s \left[\frac{\partial^2 T}{\partial x^2} + \frac{\partial^2 T}{\partial y^2} \right] = 0 \qquad \ldots (5)$$

To account for film starvation, a rigorous flow continuity equation needs to be embodied as explained in [14]. Then at a section 's' of the film, for flow continuity

$$\int_0^1 {}^s h_s\, u_s\, dz \;=\; \int_0^L h_o\, u_o\, dz \qquad \ldots (6)$$

This equation must be solved subject to the condition that the computed film width cannot exceed the actual bearing width.

Coupling of the equation set is completed by expressing the temperature dependence of kinematic viscosity via the Walther equation, then

$$\log_{10} \log_{10} \left[(v + 0.6)\right] = a \log_{10} T + b \qquad .. (7)$$

The coefficients a and b are calculated to match the lubricant characteristics and the temperature (T) is specified in Kelvin.

The solution of this equation set can be derived subject to the prescription of appropriate boundary conditions. For the generalised pressure equation the conditions of zero pressure were applied at the inlet, side and trailing edges of the film with zero gradient also prescribed at the latter when film rupture occurs. Zero pressure gradient was applied on the bearing centreline to reflect symmetry.

For the thermal solution, the boundary condition specification is more complex and is summarised in Figure 2b. These conditions reflect the experimental observation that the journal temperature (T_j) remains constant at the bulk film temperature [15] and that there is continuity of heat flux at the bush to film interface. At the bush outer surface, the temperature is assumed to increase linearly to the lobe mean film temperature (T_l).

At the lobe entry, the film temperature is assumed to vary linearly from the groove temperature (Tin) to the journal temperature while in the bush it increases from the feed temperature (Tf) to the groove temperature. The mixing mechanism taking place in the groove region which affects the prescription of lobe surface inlet temperature has been discussed in [16]. This suggests that groove temperature may be derived from a bulk mixing model where

$$Qf = Qin - Qco$$

$$Tin = \frac{QfTf + QcoTco}{Qin}$$

The temperature of the oil carried through from the upstream film (Tco) is based on the average journal and bush temperature at the film downstream face (i.e. at the plane immediately upstream of the groove). The oil carried through (Qco) and oil feed into the film (Qin) were each calculated by means of formal integration. This complex dynamic prescription has been adopted due to

its successful application to the work described in [17]. These dynamic conditions were applied to each lobe in both the lemon bore and three lobe geometries, a main objective of the present investigation is to test its applicability in profile bore geometries.

Closure of the equation set is completed by the prescription of film thickness profiles. The equations derived for these must account fully in a general way for design geometry details and loading direction effects since these depend on the attitude angle adopted by the journal.

After the first iteration, the strategic steps in the solution process may be summarised as

1. solve the generalised pressure equation
2. calculate the velocity field in readiness to solve the energy equation
3. solve the energy equation on the bearing centreplane
4. calculate the attitude angle, update boundary conditions and establish the new film axial extent
5. calculate the coefficients for the generalised pressure equation by integration over the film thickness.

This process was continued until there was pressure film convergence within 5%. On achieving this, all the bearing parameters were calculated. For power loss, the shear stress on the journal was integrated while the lubricant sideflow was derived from pressure gradient consideration and viscosity based on a local journal and bush surface temperature. This method is computationally less demanding when compared with a formal integration of the axial velocity profile. The integration was performed only where the axial film extent achieved the bearing width.

3. COMPARISON WITH OTHER PUBLISHED WORK

To confirm the numerical approach, two types of calculations were completed. Initial studies were based on the assumption of an isothermal film, while subsequent studies feature thermal behaviour. In presenting these comparisons, the isothermal work will be addressed initially.

3.1 Isothermal Analysis

The number of combinations of design parameters and operating parameters which may be selected for either the lemon bore or three lobe bearing is considerable. To restrict these, calculation has been completed for a bearing aspect ratio of 1 and for a tilt angle ϕ of $20°$ in the case of the three lobe bearing. The remaining parameters such as preset (m), eccentricity ratio (ε) and loading direction (θ) were varied in a systematic way

Table 1: Comparison of Isothermal Analysis with data from [18]
for a Lemon Bore Bearing, m = 0.5, L/D = 1

	Current Calculation			Reference 18			
ε	S	$\overline{H}$	$\overline{Q}$	S	$\overline{H}$	$\overline{Q}$	ε_ℓ
0.3	0.481	1.495	0.322	0.422	1.357	0.338	0.15
0.5	0.247	1.557	0.377	0.224	1.403	0.389	0.25
0.7	0.133	1.674	0.445	0.120	1.500	0.452	0.35
0.8	0.094	1.777	0.476	0.081	1.607	0.479	0.40
0.9	0.056	2.012	0.487	0.045	1.867	0.478	0.45

within applicable engineering limits.

The results for the lemon bore bearing are considered first under normal loading conditions (i.e. $\theta = 0$) since these can be compared with the published data in [18].

This comparison is itemised in Table 1 where it can be seen that agreement between all parameters is good.

Subsequent calculation was completed to investigate the combined influence of preset, eccentricity ratio and loading direction on the dimensionless bearing parameters. In deriving the dimensionless groups (see the nomenclature) base circle clearance (C_b) has been adopted in favour of lobe clearance (C_1) since it gives a direct measure of the clearance between the lobe inscribing circle and the journal. The results from a systematic series of calculations are shown in Figures 3 and 4.

Figure 3a illustrates the dependence of load carrying ability $(\overline{PI})$ on loading direction and preset for a fixed eccentricity ratio ε of 0.75. Clearly, it is loading direction dependent and achieves a maximum about $\theta = -20°$. Under this circumstance, the combination of loading direction and journal attitude angle gives a film profile which maximises the pressure generated and the load carrying ability consequently. This maximum does not coincide with the positioning of the rupture boundary at the downstream groove, rather it is a consequence of the rapid film thickness convergence under this operating condition. Reduction in preset systematically increases load carrying ability since the profile tends towards a cylindrical form where the load carrying capacity is best.

Figure 3b illustrates the variation in dimensionless power loss $(\overline{HI})$ which assumes a maximum for $\theta = -20°$. This is not a surprising results since the steep pressure gradients which are generated give the largest shear stress on the journal surface. When combined with the wetted surface, these result in the maximised power loss. Again, the increase in power loss with reduction in preset reflects the tendency to a near cylindrical geometry where a more substantial part of the film is thin.

Figure 3c illustrates the change in hydrodynamic sideflow $(\overline{QI})$ from the bearing which is systematic in nature with respect to preset (i.e. the tendency towards a cylindrical geometry gives a reduction due to the more conformal shape of the bush with respect to the journal). However, the minimisation of sideflow for $\theta = -20°$ is not an expected result since under this circumstance the axial pressure gradients are most severe. Close examination of the computed details revealed that the reduction in side flow under this condition is a consequence of the thin film in the region of maximum pressure gradient at the bearing edge.

Figure 4 illustrates the variation in bearing parameters under different loading directions for a range of eccentricity ratios and a preset of 0.6. This confirms the systematic variation of load carrying ability and power loss with respect to loading direction and eccentricity ratio. However, the detailed variation of hydrodynamic sideflow is not so systematic and this arises since the model accounts fully for the axial film extent which is shown in Figure 5 for both bearing types. This figure confirms that for the lemon bore profile the loaded lobe is wetted nearly completely while the unloaded lobe is only partially wetted. As expected for the case shown, the lobe wetted surfaces are interchanged for $\theta = 90°$ respectively which confirms the algorithm accuracy. For the three lobe geometry, the bearing surface is wetted nearly completely under all circumstances and thus the formal inclusion of the film wetting algorithm will not affect the prediction of performance for this bearing markedly.

Table 2: Comparison of Isothermal Analysis with [18]
for a Symmetric Three Lobe Bearing,   m = 0.5, L/D = 1

	Current Calculation			Reference 18			
ε	S	$\overline{H}$	$\overline{Q}$	S	$\overline{H}$	$\overline{Q}$	ε_1
0.3	0.386	1.578	0.137	0.360	1.478	0.147	0.15
0.5	0.183	1.711	0.159	0.180	1.574	0.165	0.25
0.7	0.090	1.963	0.183	0.085	1.794	0.189	0.35
0.8	0.059	2.183	0.200	0.054	2.014	0.208	0.40
0.9	0.033	2.594	0.231	0.034	2.290	0.232	0.45

To confirm the three lobe bearing model, predicted performance is compared with published data in Table 2 where the agreement is shown to be good.

The results of calculation for different loading directions, preset and eccentricity ratio, are shown in Figures 6 and 7. These also show that its load carrying ability $(\overline{PI})$ and power loss $(\overline{HI})$ are markedly dependent on loading direction. As discussed in [19], the load, power loss and flow $(\overline{QI})$ are maximised when the Reynolds boundary condition is established at the downstream edge of the load carrying lobe. However, the hydrodynamic flow from the film is virtually independent of loading direction as presented in [19], despite the inclusion of a formal model to establish the axial extent of the film. This arises since under the design and operating circumstances considered, the convergent film section is flooded completely, see Figure 5.

In comparing the two bearing types, the lemon bore has the best load carrying ability while the power losses are closely similar. The hydrodynamic flow requirements of the three lobe geometry exceed those of the lemon bore significantly. However, this requirement can be reduced by the inclusion of siderails as discussed in [10].

3.2   Thermal Analysis

In high speed lubrication, thermal effects are an important consideration and can define the safe operating limit under some circumstances. Before a computer model can be used with confidence, in the design role, its accuracy must be established by comparision with experiment. For this purpose, the model of the lemon bore bearing will be compared with the experimental data presented in [6] and the three lobe bearing performance will be compared with the data presented in [7] The geometry detail and physical properties adopted in each case are itemised in Tables 3 and 4 respectively.

Initial calculation for the lemon bore profile was carried out based on the assumption of ideal geometry which is free from any thermal distortion. Calculation was completed for an eccentricity ratio of 0.85 and a rotational speed of 1500 rev/min since the film is expected to be predominantly laminar under this condition. The results from this primary calculation are shown in Figure 8 and the associated predicted performance is itemised in Table 5. From Figure 8 it is clear that the agreement between measured and predicted thermal excursions is significant. The specific load itemised in Table 5 is also low in comparison with the 2MPa load imposed on the bearing. This suggests that some deformation of the bearing takes place which may assume one of two forms.

(i)   radial growth of the shaft resulting in a decrease in the nominal oil film thickness over the circumferential extent of the bearing.

(ii)   thermal distortion of the bush to give a departure from the basic film profile.

The work described in [9] suggests that both of these mechanisms are present, indeed the good agreement between experiment and theory as described in [9] was achieved by imposing the measured film thickness on the model which limits its true predictive capability. A finite element model which includes thermal distortion has been presented in the literature in [20] and which confirms the importance of this mechanism.

For the present calculation, the effect of shaft radial growth was introduced by computing its expansion according to

$$\delta = \alpha R \, (\overline{T} - To)$$

Table 3 Details of The Bearing Geometry Considered

Lemon Bore	
Journal Radius	250.0 mm
Bearing Length	500.0 mm
Radial Clearance (C_b)	0.3 mm
Bush Radial Thickness	190.0 mm
Preset (m)	0.6
Tf	45 °C
Three Lobe	
Journal Radius	37.5 mm
Bearing Length	40 mm
Radial Clearance (C_b)	0.1125 mm
Tilt Angle ϕ	0°/40°
Preset (m)	0.8
Tf	30 °C

Table 4 Thermophysical Properties

Lubricant Density	830 kg/m^3
Lubricant Specific Heat Capacity	2000 J/kg °C
Lubricant Thermal Conductivity	0.15 W/m °C
Steel Bush Thermal Conductivity	52 W/m °C
Constants in the Walther Equation	a = 3.79 b = 9.64

to give a revised journal radius

$$R^1 = R + \delta$$

This equation was integrated into the solution and a typical calculation completed, the result of which is also shown in Figure 8 and with details itemised in Table 5. Clearly the specific load carrying capacity is increased with a general rise in bearing operating temperature as opposed to a sharp increase in temperature in the loaded lobe. Therefore, this suggests that bush distortion is a more important mechanism and this was modelled by decreasing the preset (m) in a systematic manner to achieve closer agreement with measured specific load and thermal excursion. The reduction in preset gives a more circular profile and increases load carrying ability consequently. The experiment also shows clearly the cooling effect of the oil groove and this was incorporated by specifying a lobe bulk temperature at the downstream edge of the lobe as discussed in [17] with noticeable improvement in agreement. The finalised result is shown in Figure 8 and the geometry is itemised in Table 5. From Table 5, it is clear that the specific load of 2 MPa can only be achieved with a more circular geometry. Thus future developments will need to establish the distortion in a formal manner with coupling between displacement calculation and the thermal analysis.

Figure 9 illustrates the change in thermal behaviour with loading direction for the lemon bore bearing with the associated performance itemised in Table 6. The results suggest that for $0 < \theta < 90$ the thermal excursions are most extreme in the unloaded lobe. This arises due to the value of attitude angle (see Table 6) which results in a comparatively thin film in this lobe. Figure 9d illustrates significant thermal excursions in the loaded lobe with an associated load carrying ability (see Table 6). The physical mechanisms have been discussed previously under isothermal analysis, the results in Figure 9d confirm the importance with regard to thermal behaviour.

Calculation was also completed for the three lobe bearing for an eccentricity ratio of 0.75 and a rotational speed of 10,000 rev/min. The results from this series of calculation were compared with the experimental data from the comprehensive experiments reported in [10]. The comparisons are presented in Figures 10 and 11 for the nominally equivalent loads shown inset. These comparisons are presented for the boundary conditions which reflect no heat transfer between the groove and the downstream edge of the preceding lobe. This does not lead to good agreement between the measured and predicted thermal behaviour in the symmetric geometry where thermal effects are more significant and which has been discussed in detail in [9].

Table 5 Calculation Conditions for Lemon Bore Thermal Analysis

Model Number	Specific Load (MPa)	Journal Torque (Nm)	Side flow (l/s)	Altitude Angle	Comment
1	0.366	258.0	2.81	84	$m = 0.6$; $\varepsilon = 0.85$ $R = 250.00$ mm
2	0.88	277.8	1.56	84	$m = 0.9$; $\varepsilon = 0.9$ $To = 25$; $\alpha = 12\ 10^{-6}$ $R = 250.13$ mm*
3	2.08	435.1	2.7	51	$m = 0.4$; $\varepsilon = 0.85$ $R = 250.00$

+ See Figure 8 for thermal excursion
* Final converged journal radius

Table 6 Loading Direction Effects on a Rigid Lemon Bore Bearing; $m = 0.6$, $\varepsilon = 0.85$

Loading Direction (θ)	90	45	0	-45
Load Capacity (N)	44952	39677	92049	147150
Journal Torque (Nm)	267.53	254.66	258.35	309.67
Flow rate (l/s)	2.84	2.86	2.81	2.63
Attitude Angle	21	60	85	8

For this model, the agreement with measurement for the tilted geometry is improved since the film is generally convergent in the predominant loaded lobe under most circumstances. However, the model did not capture the heat transfer mechanism which is also evident for the tilted bearing geometry and thus boundary conditions were introduced as explained in [17] and which were applied to the lemon bore geometry as explained in the preceding paragraphs.

The results of these further calculations are shown in Figures 12 and 13 for generally equivalent loads to those used in Figures 10 and 11. With the revised boundary condition, the agreement between prediction and measurement is much improved for each bearing geometry for the range of loading conditions investigated. This confirms the importance of the heat transfer mechanism at the feed groove which has also been discussed in [21] and lend support to the accuracy of the predictive model. No thermal distortion complexities were observed in this case since the thermal excursions are not so severe and journal radial growth and bush displacement will be small consequently.

4. CONCLUSIONS

Finite element models to predict the behaviour of lemon and three lobe bearings subjected to different loading directions has been presented and compared with other published data and experimental results. The models agree closely with published isothermal analysis and it is confirmed that including wetted film extent when predicting bearing performance is not of significant importance for these bearing types. The results show that loading direction effects are not dramatic in lemon bore profiles, but lead to significant changes in three lobe geometries.

With regard to thermal behaviour, comparison with results for large lemon bore geometries confirms the need to include distortion analysis particularly under severe thermal conditions. This is in agreement with findings described in [9]. Also, for lemon bore geometry, loading direction does have a significant effect on load carrying ability (which will be reflected in film thickness for a prescribed load application).

For the three lobe bearings considered, the quality of agreement between prediction and measurement is improved with the inclusion of a heat transfer mechanism at the groove. This has been included as a boundary condition specification, but the physics requires a more formal and rigorous investigation with a consideration of groove detail geometry.

REFERENCES

1. Garner, D.R., Lee, C.S. and Martin, F.A.; Stability of Profile Bore Bearings : Influence of Bearing Type Selection; Tribology International, 13, 1980, p204-210.

2. Albin, F.T., Campbell, J. and Garner, D.R.; The Technical Development and Market Exploitation of Novel Manufacturing Techniques for High Speed Plain Bearings; Proc. I.Mech.E., 200, 1986, p77-83.

3. Martin, F.A. and Ruddy, A.V.; The Effect of Manufacturing Tolerances on the Stability of Profile Bore Bearings; I.Mech.E. Conference. Vibrations in Rotating Machinery, York, Sept. 1984, p287-293, M.E.P.

4. Leader, M.E., Flack, R.D. and Lewis, D.W.; Experiments on the Stability and Response of a Flexible Rotor in Three Types of Journal Bearings; Trans. ASLE 25, 1982, p280-298.

5. Gardner, W.W. and Ulschmid, J.G.; Turbulence Effects in Two Journal Bearing Applications; Trans. ASME (JOLT), 96, 1974, p15-21.

6. Hopf, G. and Schuler, D.; Investigations on Large Turbine Bearings Working Under Transitional Conditions between Laminar and Turbulent Flow; Trans. ASME (JOT), 111, 1989, P628-634.

7. Basri, S.B. and Gethin, D.T.; An Experimental Investigation into the Thermal Behaviour of a Three Lobe Profile Bore Bearing; (to be published in Trans. ASME (JOT).

8. Safar, L. and Szeri, A.Z.; Thermohydrodynamic Lubrication in Laminar and Turbulent Regimes; Trans. ASME (JOLT), 96, 1974, p48-57.

9. Mittwollen, N. and Glienicke, J.; Operating Conditions of Multi-Lobe Journal Bearings Under High Thermal Loads; Trans. ASME (JOT), 112, 1190, p330-340.

10. Gethin, D.T. and Basri, S.B.; An Experimental and Numerical Investigation into the Thermal Behaviour of a Three Lobe Profile Bearing; Proc. I.Mech.E. (c), 205, 1991, p251-264.

11. Gethin, D.T. and El-Deihi, M.K.I.; Thermal Model for a Twin Axial Groove Bearing Subjected to a Varying Loading Direction and its Verification; Tribology International, 24, 1991, p131-136.

12. Coles, J.A. and Hughes, C.I.; Oil Flow and Film Extent in Complete Journal Bearings; Proc. I.Mech.E., 170, 1956, p499.

13. Dowson, D.; A Generalised Reynolds Equation for Fluid Film Lubrication; International Journal of Mechanical Engineering Sciences; 4, 1962. p159-170.

14. Gethin, D.T. and Medwell, J.O.; Analysis of High Speed Bearings Operating with Incomplete Films; Tribology International, 18, p340-346.

15. Dowson, D.; Hudson, J.D.; Hunter B. and March, C.N.; An Experimental Investigation of the Thermal Equilibrium of Steadily Loaded Journal Bearings; Proc. I.Mech.E., 181, 1966-68, p70.

16. Heshmat, H. and Pinkus, O.; Mixing Inlet Temperatures in Hydrodynamic Bearings; Trans. ASME (JOT), 108, 1986, P231-248.

17. El-Deihi, M.K.I. and Gethin, D.T.; A Thermohydrodynamic Analysis of a Twin Axial Groove Bearing Under Different Loading Directions and Comparison with Experiment; Trans. ASME (JOT), 114, 1992, p304-310.

18. Lund, J.W. and Thomsen, K.K.; A Calculation Method and Data for the Dynamic Coefficients of Oil Lubricated Journal Bearings; Topics in Fluid Film Bearing and Rotor Bearing System Design and Optimisation; ASME Publication 100118, New York, 1978.

19. Gethin, D.T.; Thermodynamic Behaviour of Profile Bore Bearings; Proc. I.Mech.E., Tribology - Friction, Lubrication and Wear, 50 Years On, 1987, p145-156.

20. Gethin, D.T.; An Investigation into Plain Journal Bearing Behaviour Including Thermoelastic Deformation of the Bush; Proc. I.Mech.E. (c), 199, 1985, p215-223.

21. Knight, J.D. and Gadimi, P.; Effects of Modified Effective Length Models of the Rupture Zone on the Analysis of a Fluid Film Journal Bearing; STLE Preprint 91-AM-5C-1.

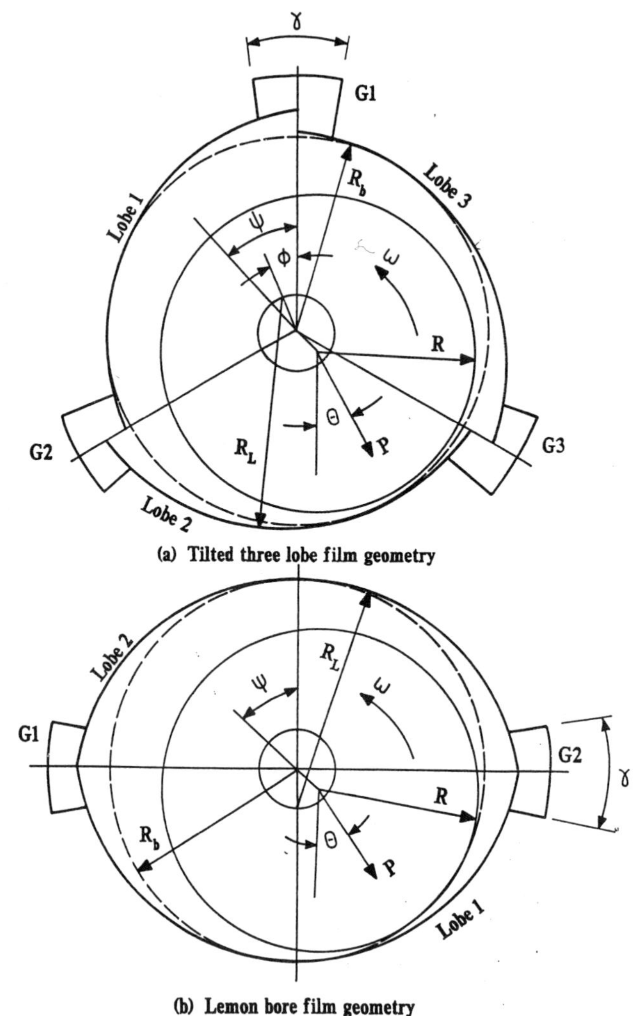

(a) Tilted three lobe film geometry

(b) Lemon bore film geometry

Fig 1   General arrangement of the profile bore bearings

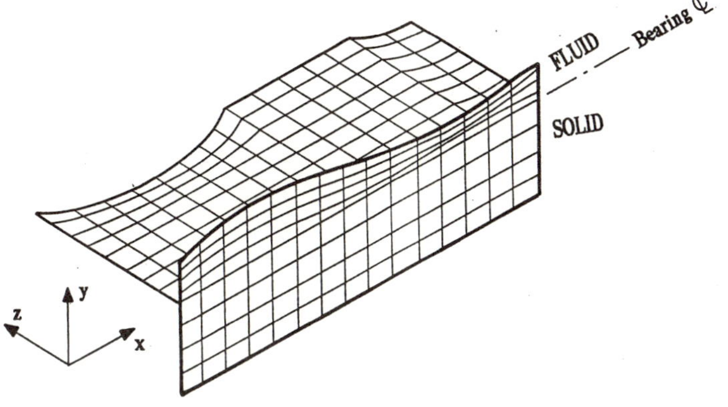

(a) Finite element mesh for hydraulic and thermal analysis

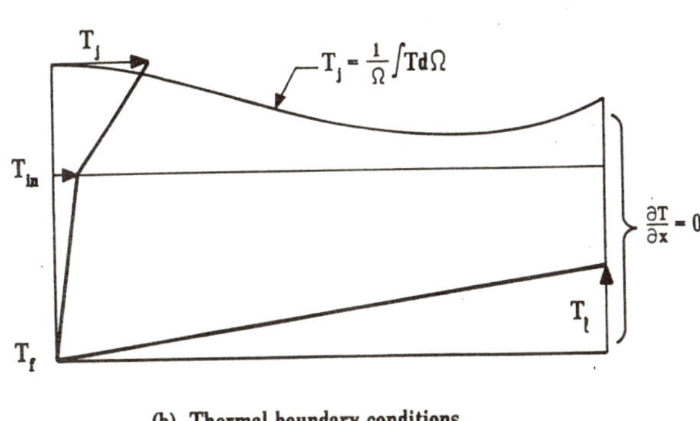

(b) Thermal boundary conditions

Fig 2   Finite element mesh and thermal boundary conditions

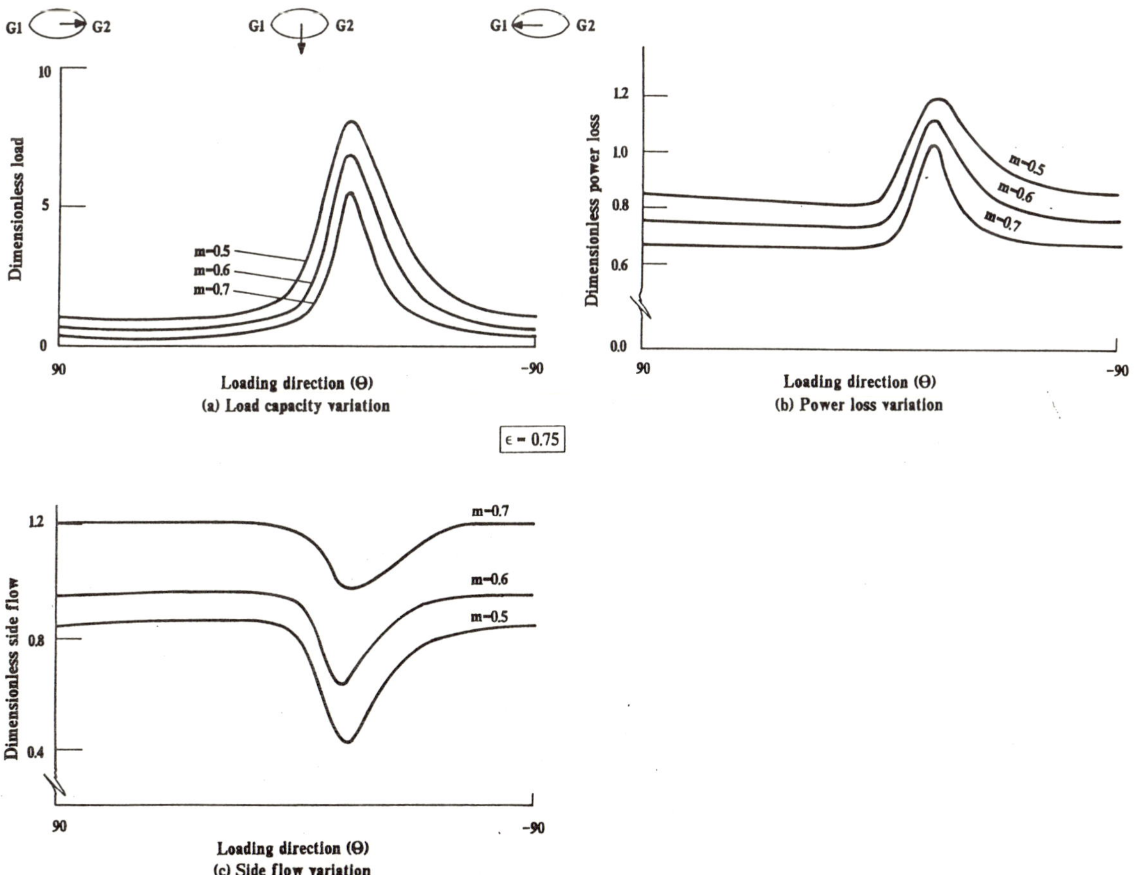

Fig 3   Lemon bore bearing performance for different preset
and loading direction; $L/D = 1.0$

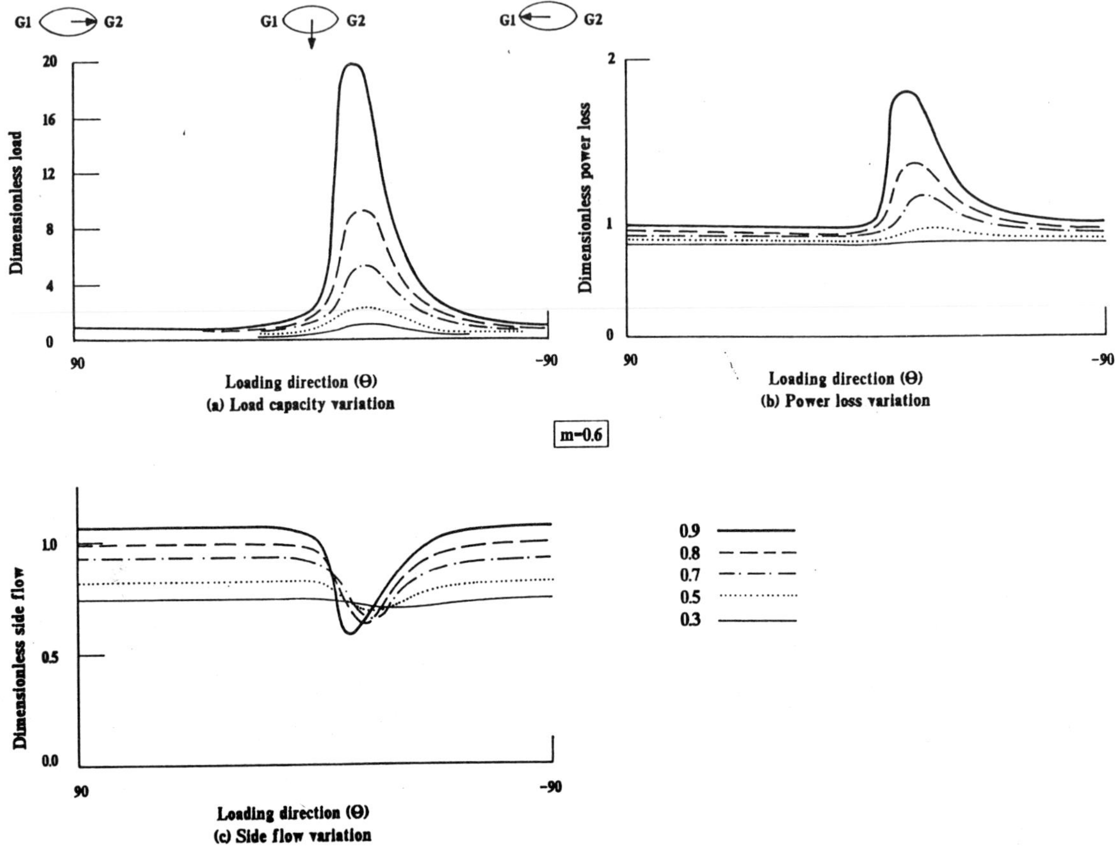

Fig 4   Lemon bore bearing performance for different eccentricity ratio and loading direction; L/D = 1.0

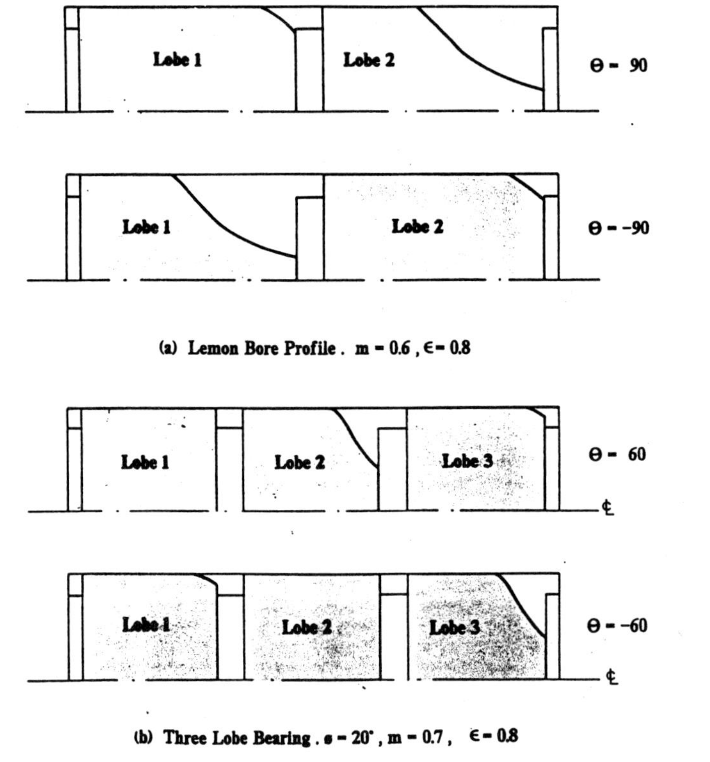

Fig 5   Computed wetted film extent under different loading directions

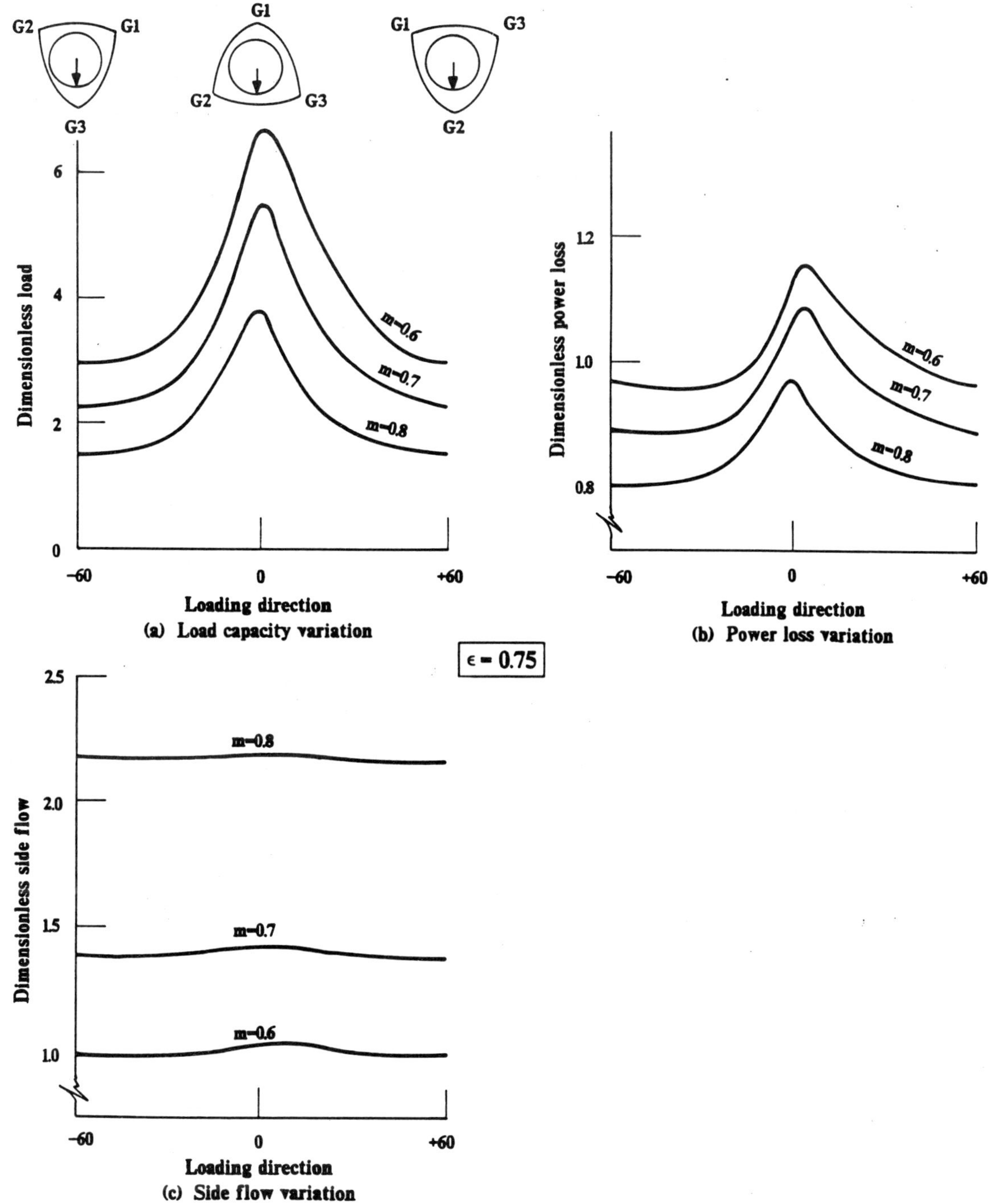

Fig 6   Symmetric three lobe bearing performance characteristic for different preset loading direction; L/D = 1.0

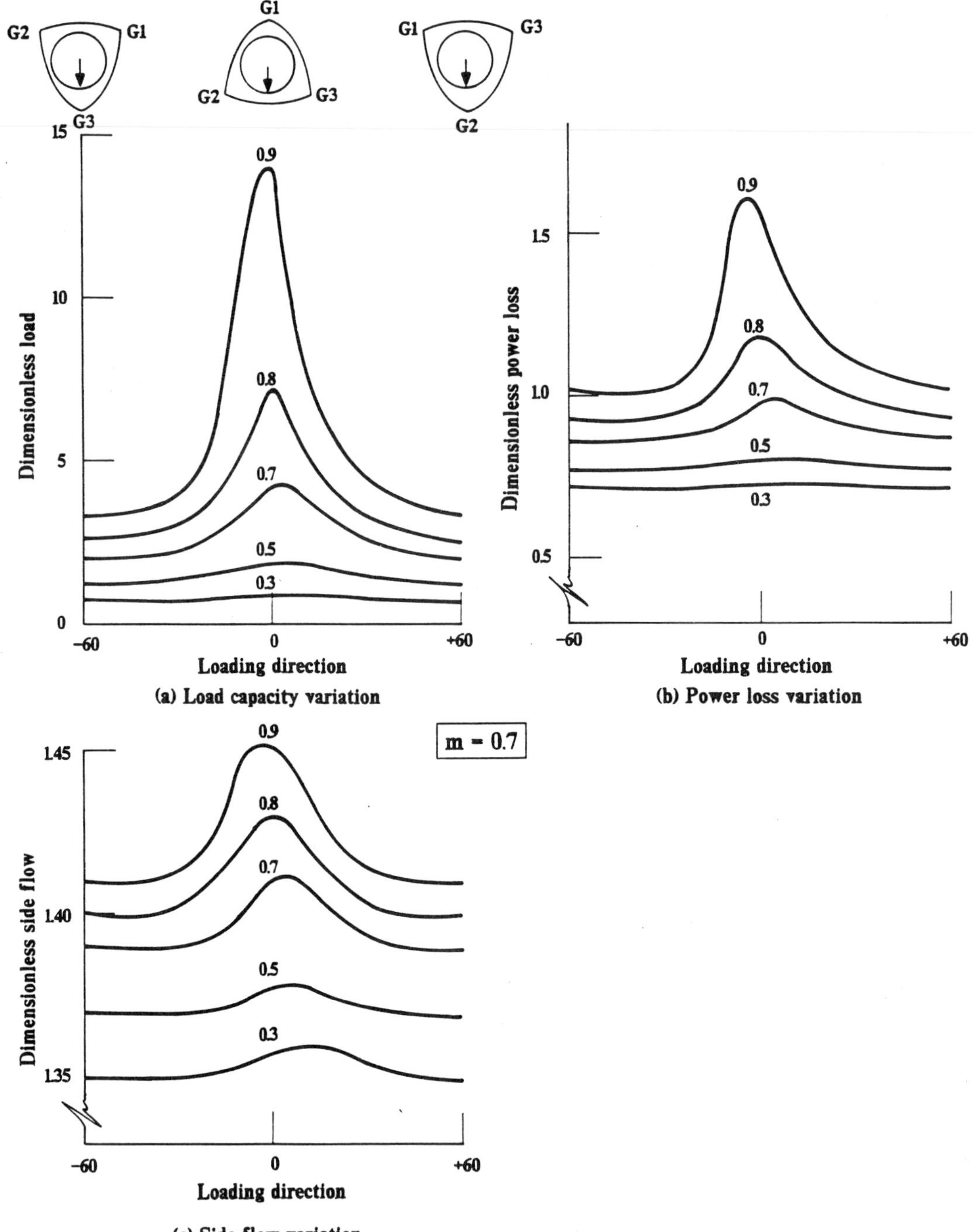

Fig 7   Symmetric three lobe bearing global behaviour for different eccentricity ratios and loading direction; L/D = 1.0

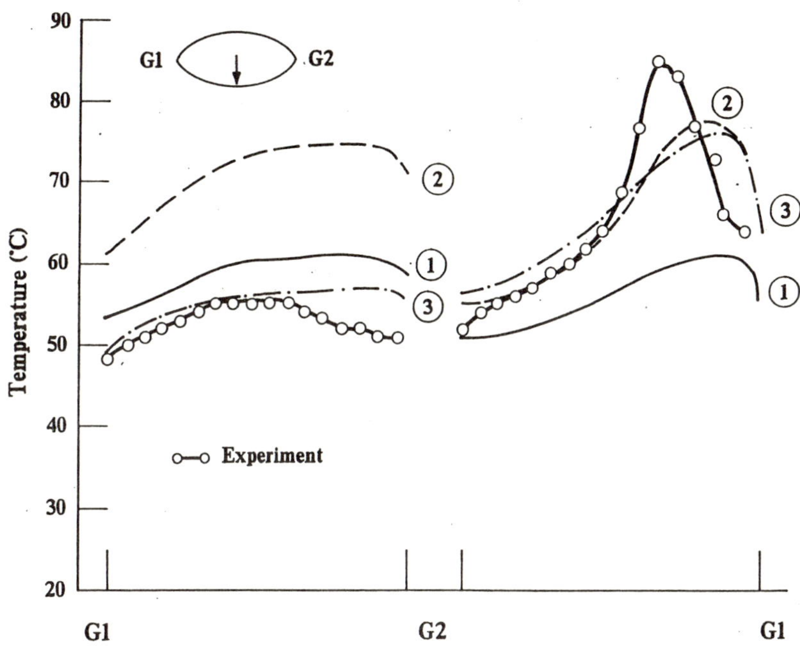

Fig 8   Comparison of thermal models with experimental
         data for a lemon bore bearing; ω = 1500 rev/min

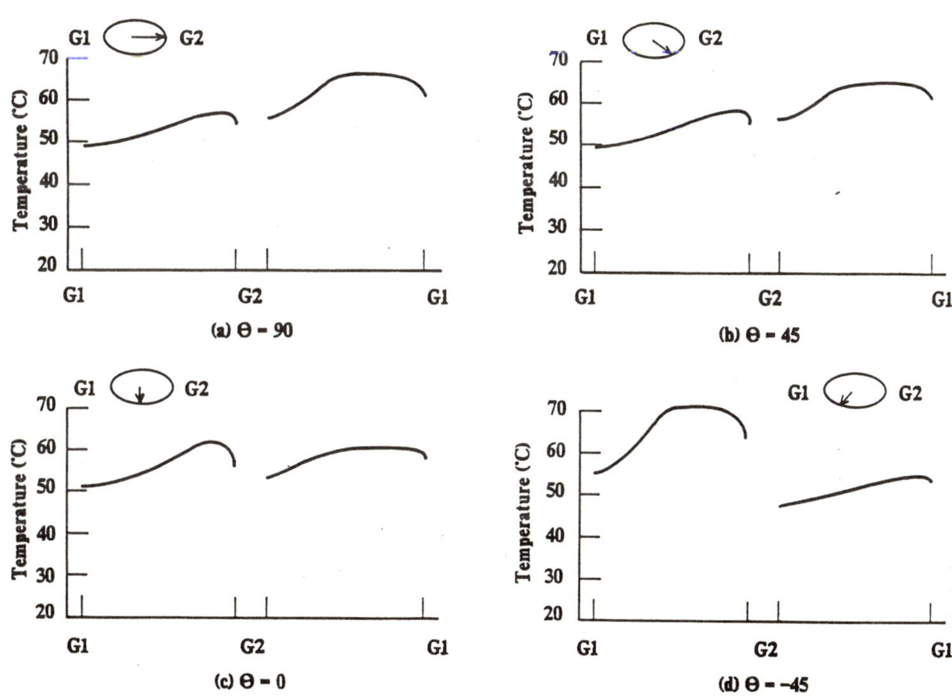

Fig 9   The effect of loading direction on lemon bore
         bearing thermal behaviour; m = 0.6, ε = 0.85,
         ω = 1,500 rev/min, R = 250.0 mm, L/D = 1.0,
         C_b/R = 0.003

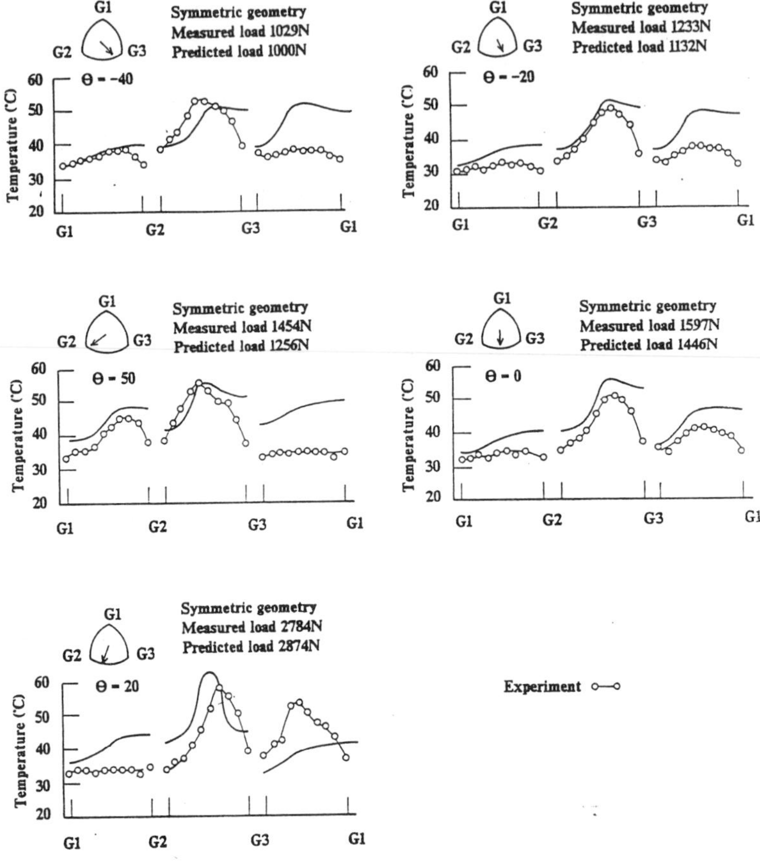

Fig 10   Comparison between predicted and measured performance for a symmetric three lobe bearing; m = 0.8, ε = 0.75, ω = 10,000 rev/min, R = 37.5 mm, L/D = 0.53, C_b/R = 0.003

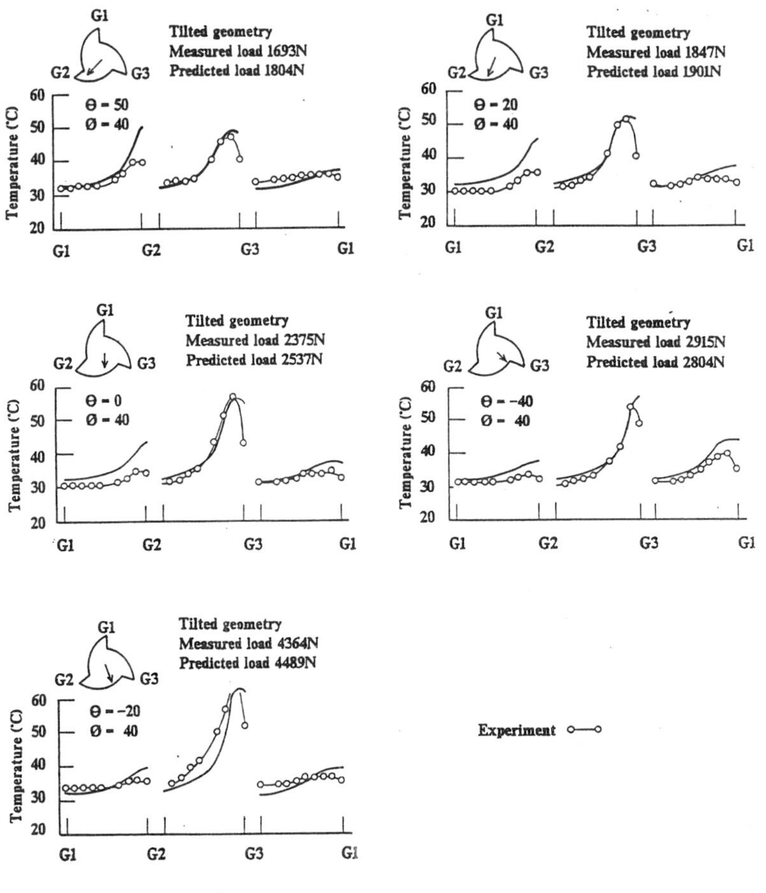

Fig 11   Comparisons for a tilted three lobe bearing; m = 0.8, ε = 0.75, ω = 10,000 rev/min, R = 37.5 mm, L/D = 0.53, C_b/R = 0.003

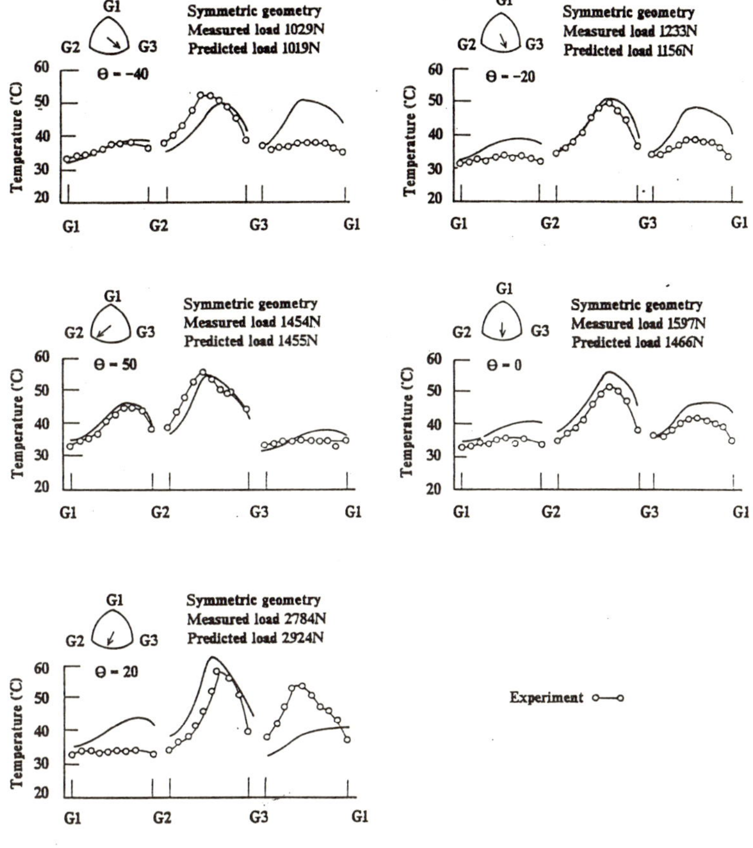

Fig 12   Comparisons including groove heat transfer model for a symmetric bearing; $m = 0.8$, $\varepsilon = 0.75$, $\omega = 10{,}000$ rev/min, $R = 37.5$ mm, $L/D = 0.53$, $C_b/R = 0.003$

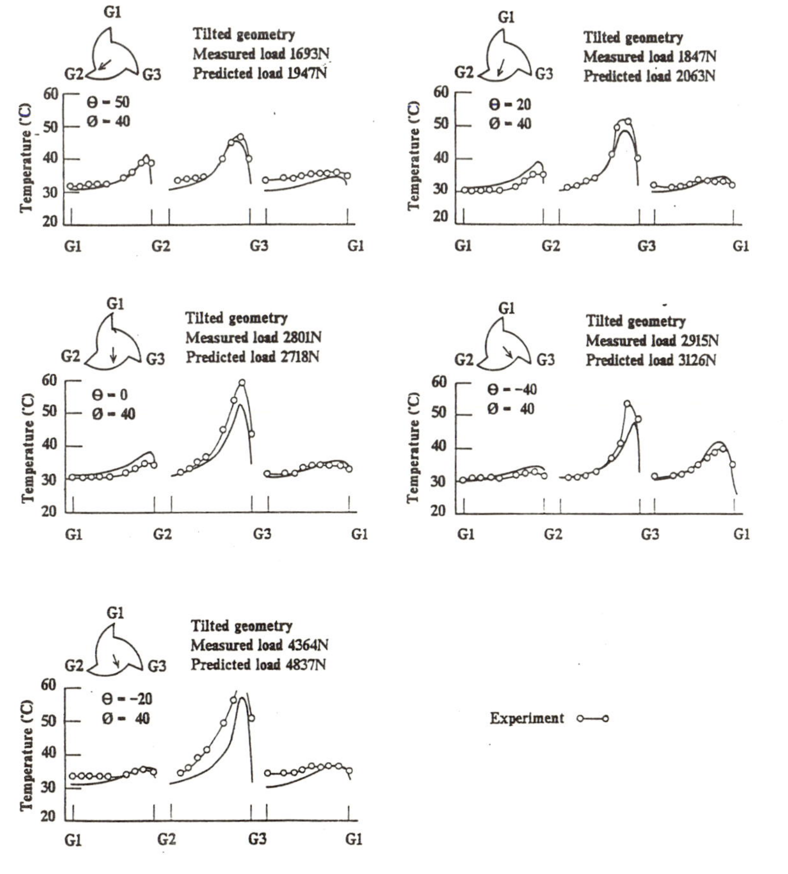

Fig 13   Comparisons including groove heat transfer for a tilted three lobe bearing; $m = 0.8$, $\varepsilon = 0.75$, $\omega = 10{,}000$ rev/min, $R = 37.5$ mm, $L/D = 0.53$, $C_b/R = 0.003$

A design study of partially grooved steadily loaded journal bearings

C K K LAI, BEng, AMIMechE and C M TAYLOR, BSc, MSc, PhD, DEng, CEng, FIMechE
The Institute of Tribology, Department of Mechanical Engineering, University of Leeds, UK

Synopsis

Partial central circumferential groove journal bearings are used widely in the internal combustion engine. The design procedures for such bearings are limited to the extent that a number of important aspects do not in general receive attention. For instance, analytical approaches do not normally consider the influence of the load direction with relation to the groove position. In addition lubricant flow continuity in the film is not ensured. As a precursor to the examination of such effects in dynamically loaded bearings, a study of steadily loaded situations has been undertaken.

This paper will present design data in dimensionless form for a 180° partially grooved plain journal bearing taking into account the aspects identified above. A range of dimensionless supply pressure, bearing width-to-diameter ratio and eccentricity ratio were considered, and bearing load capacity, oil flow rate and power loss data are presented. As well as considering the influence of load direction upon bearing performance, the effect of oil groove extent was also examined. The performance characteristics exhibit some interesting contrasts with those for a full central circumferential groove bearing and the physical interpretation of these differences will be explored.

Notation

a	Groove width
b	Bearing width
c_d	Bearing diametral clearance
c	Bearing radial clearance
d	Bearing diameter
b/d	Width-to-diameter ratio
e	Eccentricity
H	Power loss
h	Oil film thickness
h_{cav}	Film thickness at rupture boundary
N	Rotational frequency
p	Oil film pressure
p_f	Oil feed pressure
Q	Oil flow rate
Q_s	Bearing side leakage flow rate (two sides)
Q_v	Velocity induced flow
Q_f	Feed pressure flow
W	Load
η	Lubricant dynamic viscosity
ψ	Attitude angle
ε	Eccentricity ratio (e/c)
$\overline{\ \ }$	A bar on top of a variable represents a dimensionless quantity.

Dimensionless Groups

$$\overline{p} = \frac{(c_d / d)^2}{\eta N} p \qquad \overline{W} = \frac{(c_d / d)^2}{\eta Nbd} W$$

$$\overline{Q} = \frac{1}{bdNc_d} Q \qquad \overline{H} = \frac{(c_d / d)}{\eta N^2 bd^2} H$$

1 Introduction

The influence of lubricant supply arrangements on the performance of plain journal bearings has attracted significant attention over the years. Many challenging problems remain to be addressed in order to effect an improved understanding and to further the sensitivity of design procedures to the impact of the geometry and location of supply holes and grooves.

The availability of rationalized and validated plain bearing design procedures incorporating a consideration of supply arrangement issues is limited. For steady operating situations (or applications with loads and speeds which vary slowly) where axial grooves are called for, reference [1] presents a comprehensive design process which utilizes empirical data. Similarly design documentation has been published for cases where a bearing may experience a range of different effective steady state circumstances and for which a full central circumferential groove is appropriate [2]. However, for situations with a supply hole or partial groove and for dynamically loaded bearings precious little reliable information is accessible.

In the present paper the performance characteristics of steadily loaded, partially grooved, plain journal bearings are explored with a particular consideration of the physical interpretation of the predicted behaviour. A range of design data is also presented. The basic configuration which has been examined is shown in figure 1. A 180°, central, partial groove is considered and the steady load is taken to bisect the groove extent and to be directed towards the ungrooved portion of the bush. In undertaking the analysis for this situation full consideration of lubricant flow continuity has been ensured with a detailed investigation of film reformation and the true extent of the cavitated region. The investigation has been extended by a consideration of the influence of load direction with a 180° partial groove and a study of the effect of groove extent with the load direction central with respect to the groove.

The research reported formed part of a study leading towards the development of an improved analytical approach to the prediction of the performance of dynamically loaded, internal combustion engine bearings with partial grooving.

2 Analysis

2.1 Reynolds Equation and Boundary Conditions

A common form of Reynolds equation for a steadily loaded journal bearing, with an iso-viscous, incompressible lubricant can be expressed as;

$$\frac{\partial}{\partial x}\left(\frac{h^3}{12\eta}\frac{\partial p}{\partial x}\right) + \frac{\partial}{\partial y}\left(\frac{h^3}{12\eta}\frac{\partial p}{\partial y}\right) = \frac{U}{2}\frac{\partial h}{\partial x} \qquad (1)$$

where x and y are the circumferential and axial coordinates of a developed bearing area.

Over the years, various physical models have been proposed for the determination of the cavitated region boundaries. A summary of the work carried out in this field has been presented by Dowson and Taylor [3]. The "Reynolds" and "Jakobsson and Floberg" boundary conditions are generally accepted as the most realistic boundary conditions for the rupture and reformation boundaries respectively. By applying the principle of flow continuity across the cavitation boundary interfaces, it can be shown that;

Rupture boundary (Reynolds boundary conditions)
$$p = 0 \qquad (2)$$

$$\frac{\partial p}{\partial x} = \frac{\partial p}{\partial y} = 0 \qquad (3)$$

Reformation boundary (Jakobsson and Floberg boundary conditions [4])
$$p = 0 \qquad (4)$$

$$\frac{Uh_{cav}}{2} = \frac{Uh}{2} - \frac{h^3}{12\eta}\left(\frac{\partial p}{\partial x} - \frac{\partial p}{\partial y}\frac{\partial x}{\partial y}\right) \qquad (5)$$

The above reformation boundary condition (equation 5) describes the relationship between the local film thickness, the pressure gradients and the fractional width of the recirculating oil. In simple terms, this expression implies that the oil flow passing through the reformation boundary must balance to the flow passing through the rupture boundary.

2.2 Computational Method

The finite difference method was employed to approximate the Reynolds equation. The "Reynolds" and "Jakobsson and Floberg" boundary conditions were indirectly imposed using the cavitation algorithm presented by Rowe and Chong [5]. The use of the finite difference approximation method resulted in a system of simultaneous equations. The Tri-Diagonal Matrix Algorithm solution technique was adopted [6].

After obtaining a converged solution of the pressure field, the hydrodynamic forces components were computed and examined. It was necessary to ensure that correct alignment between the load vector and the stationary bearing shell (partially grooved) was achieved. If this was not the case, the attitude angle was updated and the pressure field iteration was repeated until both the pressure and attitude angle convergence criteria were satisfied.

2.3 Range of variables considered

Computations were carried out for the following range of independent variables though only a limited presentation of results can be made here. Further details may be found in reference [7].

- Non-dimensional supply pressure $(\overline{p}_f) = 0.05$, 0.1, 0.5, 1.0, 5.0
- Bearing width-to-diameter ratio (b/d) = 0.2 - 1.0 in steps of 0.2, plus 0.5
- Eccentricity ratio $(\varepsilon) = 0.2 - 0.9$ in steps of 0.1, plus 0.95
- Groove width to bearing width ratio (a/b) = 0.2
- Groove located at bearing angle 0°- 180° (except otherwise stated, figure 5 and 6)
- Load direction - vertically downwards (except otherwise stated, figure 7)

3 Influence of Supply Pressure on Partial Groove Bearing

Figure 2 is a plot of Dimensionless Load versus the Ratio of Supply Pressure to Specific Load of a partially

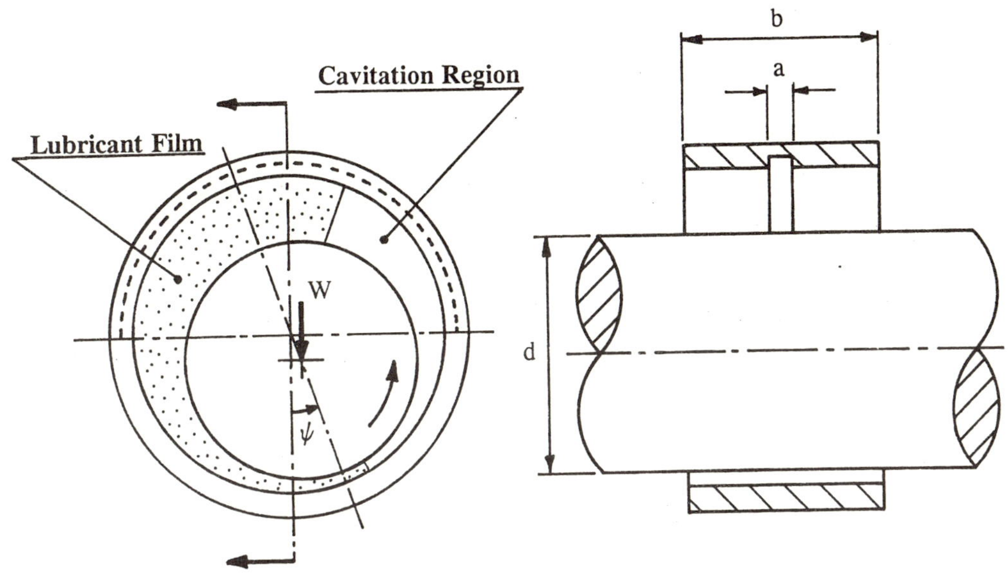

Figure 1 Sketch of a 180° Partial Groove Journal Bearing

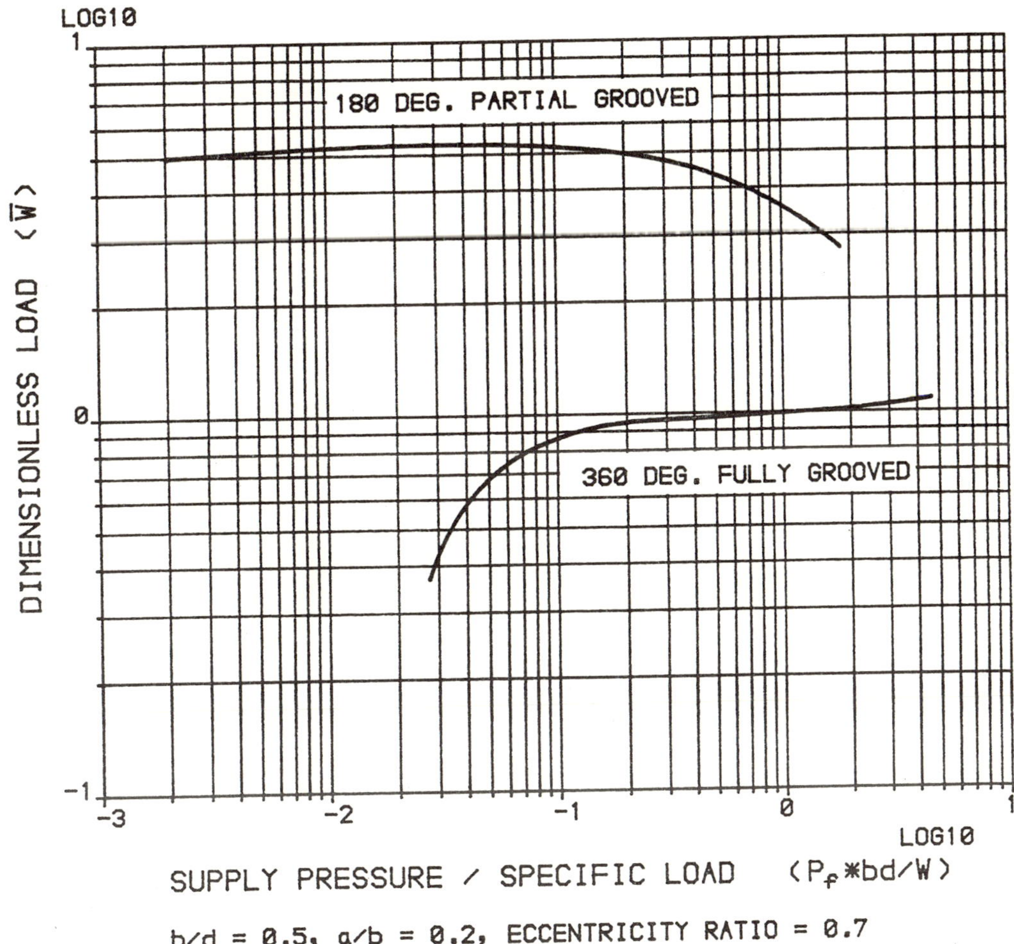

b/d = 0.5, a/b = 0.2, ECCENTRICITY RATIO = 0.7

FIGURE (2) DIMENSIONLESS LOAD VERSUS

SUPPLY PRESSURE / SPECIFIC LOAD

grooved and a full central circumferential grooved bearing. For a given bearing geometry, dynamic viscosity and shaft rotational frequency, an increase of supply pressure will cause the region of cavitation to decrease. A similar presentation for a fully grooved bearing has been given by Clayton and Taylor [8]. In the case of a partially grooved bearing, the change of the cavitation region for different supply pressures is shown in figure 3.

For the fully grooved bearing, at a very low supply pressure, oil starvation will occur and result in a low load capacity [8]. As the supply pressure is gradually increased, the cavitation region will decrease and effect an increase in load capacity. Eventually, the hydrodynamic load capacity will reach a limit when the cavitation region has totally vanished. A further increase of supply pressure will not cause the load capacity to increase as the supply pressure will not contribute to the load capacity for a complete oil film with a full central circumferential groove.

In the case of a partially grooved bearing, providing that the groove is not in the loaded region (where the hydrodynamic pressure would be high), the bearing will operate at a very low oil supply pressure, even with a zero supply pressure. This is because the feeding of the lubricant does not totally rely on the supply pressure. Lubricant will be entrained into the bearing via the groove end to form the load carrying film. This entrainment is caused by the relative motion of the bounding surfaces. Figure 3 gives a clear picture of film extent for different supply pressures. It can be seen that for zero supply pressure, lubricant flows into the bearing via the groove end only. As the supply pressure gradually increases, the cavitation region decreases, which is to be expected. Unlike the fully grooved case which relies on the feed pressure to force lubricant into the bearing clearance space, provided that the groove is properly located and has a reasonable width compared to the bearing width, sufficient lubricant will be dragged into the bearing to form the load carrying film. A partially grooved bearing not only has a better load carrying characteristic when compared with a fully grooved bearing (figure 2), but the entraining of the lubricant into the bearing to form the load carrying film is a reliable and effective feeding mechanism.

Although an increase of supply pressure will cause a decrease in the cavitation region of the partially grooved bearing, there will only be a small increase in the load carrying capacity as can be seen in figure (2) and deduced from figure (3). This can be interpreted as implying that the increase in supply pressure only marginally influences the main load bearing portion of the film. With the partially grooved bearing in question, a 180° groove has been located on the top half of the bearing. When the supply pressure is increased to a certain limit, the role of the hydrostatic pressure in the groove becomes important. This is because the hydrostatic load acts counter to the generated hydrodynamic load. Such an adverse effect is clearly

shown in figure 2. However, it must be emphasized that a suitable increase of supply pressure will improve the thermal characteristics and hence reduce the possibility of difficulties due to overheating.

4 Dimensionless Design Data for a 180° partial Groove Steadily Loaded Journal Bearing

Figures 4a,b,c and d are the generated design data for a 180° partially grooved steadily loaded journal bearing. In figure 4a, an effective comparison of load capacity can be made for dimensionless supply pressures equal to 0.05 and 5.0. At high supply pressure, the load capacity dropped rapidly as the eccentricity ratio decreased. The significance of the ratio of dimensionless supply pressure to bearing specific load has already been shown in figure 2. As the supply pressure became too high when compared with the generated hydrodynamic pressure, a very low load carrying capacity resulted. For extreme cases, a zero or even negative load capacity was encountered. This meant that a load had to be applied in the reverse direction so that the shaft could be located at that particular eccentricity. Such data were considered to be of no practical significance. For this reason, the data corresponding to negligible load carrying capacity were not included in the results presented. This explains why in figures 4a - 4d the curves do not have the same terminal point, particularly at low eccentricity ratio and low width-to-diameter ratio.

Figure 4b is a plot of the shaft centre polar locus, for the range of bearing width-to-diameter ratios and supply pressures considered. At low eccentricity ratio, the attitude angle is more affected by the supply pressure. This is due to the increase of supply pressure causing the cavitation region to decrease. This results in a change in load capacity and the bearing equilibrium position. At high eccentricity, the effect of supply pressure on journal position is small. This is because the magnitude of the generated hydrodynamic force is substantial when compared with the effects due to an increase of supply pressure.

Figure 4c shows the dimensionless power loss as a function of the eccentricity ratio for a series of bearing width-to-diameter ratios. The increase of supply pressure causes the cavitation region to decrease. More lubricant is under shearing action and hence a higher power loss results. However, such an increase in power loss is small. The results presented show that, at an eccentricity ratio of 0.8, when the dimensionless supply pressure is increased from 0.05 to 5.0, there is only a corresponding 12% to 17% increase in power loss for the range of bearing width-to-diameter ratios considered.

Figure 4d shows a range of dimensionless flow rate versus eccentricity ratio for the range of bearing width-to-diameter ratio and dimensionless supply pressures previously considered. In general, for a given eccentricity ratio, the observed trend shows that the oil flow rate increases as the bearing width-to-diameter

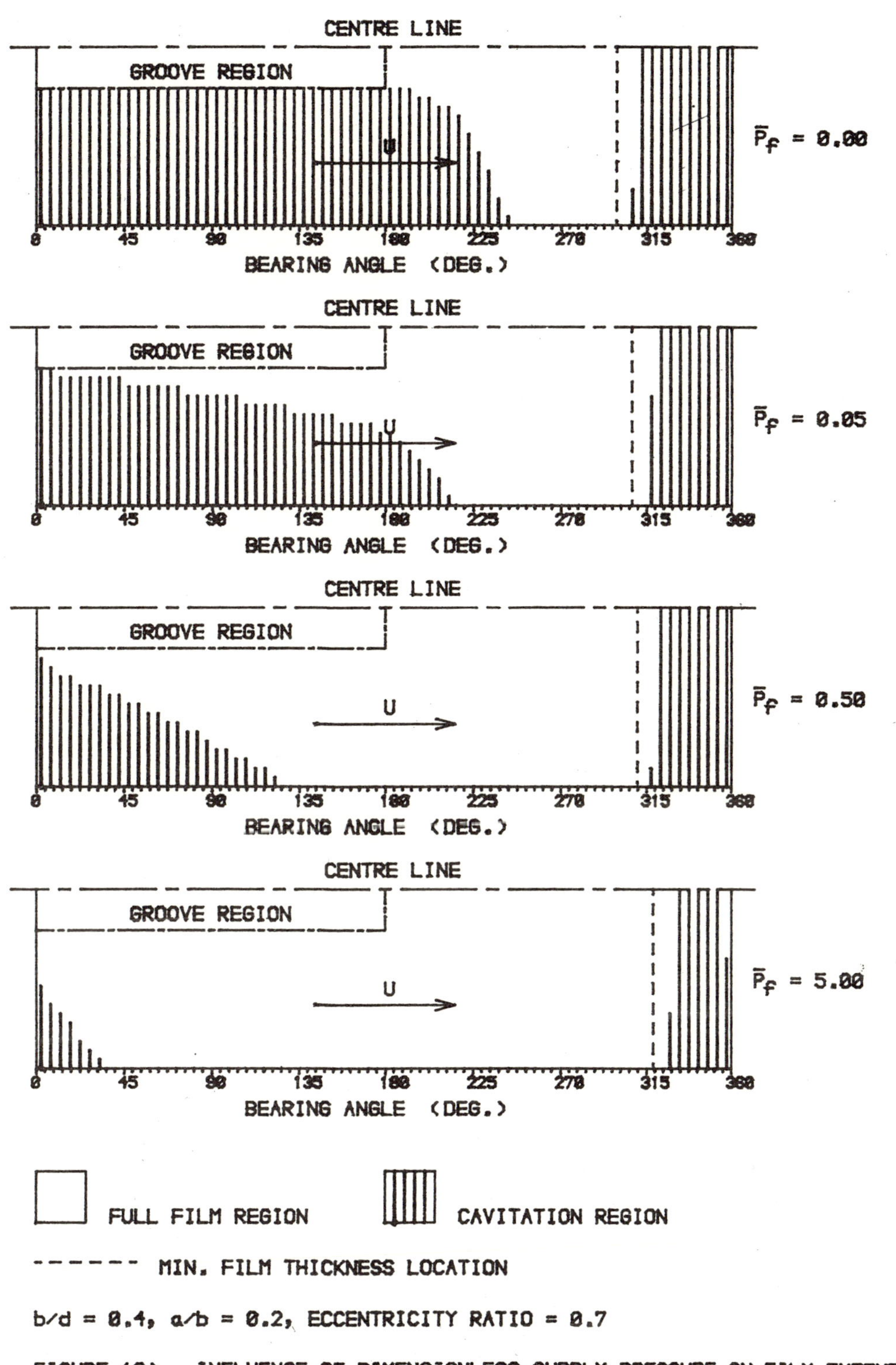

FIGURE (3)   INFLUENCE OF DIMENSIONLESS SUPPLY PRESSURE ON FILM EXTENT

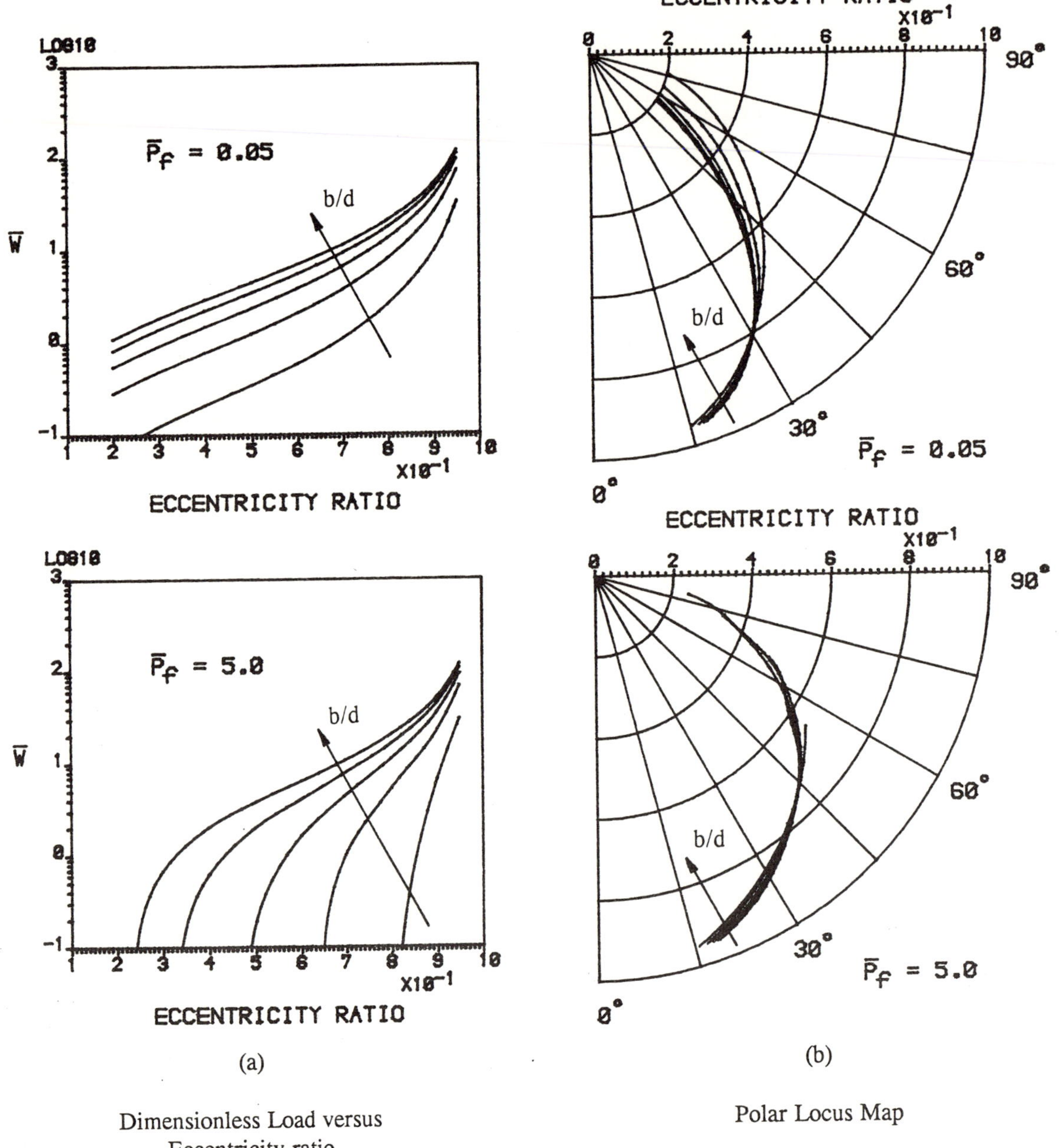

(a)

Dimensionless Load versus
Eccentricity ratio

(b)

Polar Locus Map

Figure 4 Dimensionless Design Data of a 180° Partial Groove Journal Bearing.

(b/d : 0.2 in steps of 0.2 to 1.0)

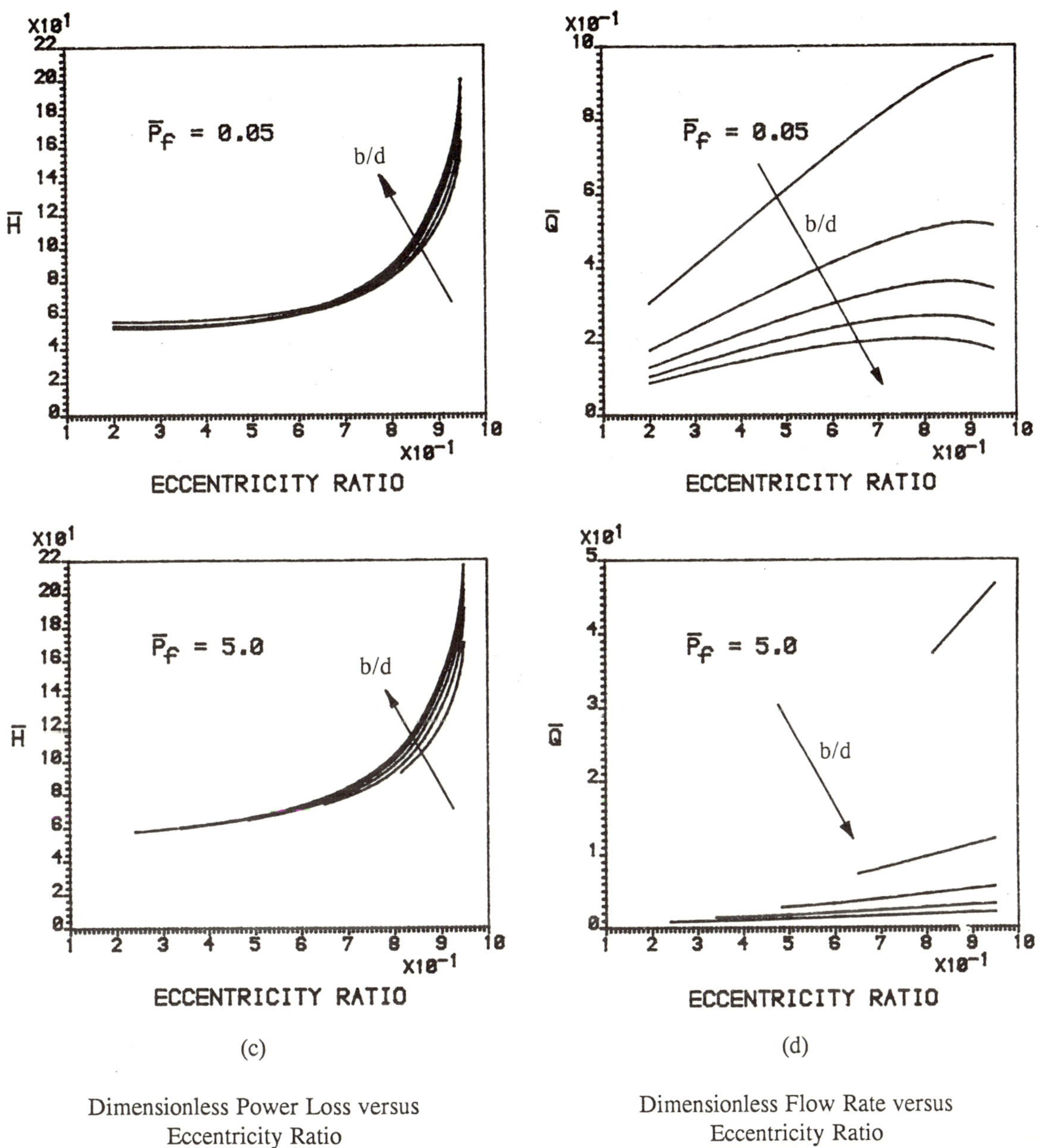

(c)

Dimensionless Power Loss versus
Eccentricity Ratio

(d)

Dimensionless Flow Rate versus
Eccentricity Ratio

Figure 4 (continued)

ratio decreases, which is expected.

A range of design data for the 180° partial grooved steadily loaded journal bearing has been presented. However, there are numerous alternative designs of such partially grooved bearings and it is not possible to consider the full spectrum. The purpose of this presentation has been to present a physical appreciation of partially grooved bearing performance as an aid to understanding and assistance in design.

5 Effect of Load Direction on the Performance of a 180° Partially Grooved steadily loaded journal bearing.

Figure 5 presents the influence of load direction on the predicted performance of a 180° partially grooved bearing. Figure 6 is a plot of film extents corresponding to different load directions.

As expected, the direction of load has a great influence on partially grooved bearing performance. With the 180° partially grooved bearing in question, at eccentricity ratios of 0.7 and 0.9, it was noted that there was approximately a factor of 4.5 difference in load capacity when the load was acting on the grooved and the ungrooved parts. A rapid change in load capacity was observed when the load was directed near the groove ends. This effect was found to be more pronounced at high eccentricity ratio. The large variation in load capacity can be explained with reference to figure 6. Since the major load carrying oil film region is in the vicinity of the minimum film thickness location, the angular location of this region depends on the load direction. When the load carrying film is in the grooved region, the area available to generate the hydrodynamic pressures is small when compared with the ungrooved region. In addition, as the high pressure load carrying film is close to the groove, lubricant will be forced back into to the groove and further the high pressure region is disturbed by the groove. It is therefore expected that a higher load capacity can be obtained when the load is arranged to act on the ungrooved part of the bearing.

A similar characteristic was observed for the power loss. For the load acting on the grooved and the ungrooved parts of the bearing, at eccentricity ratios of 0.7 and 0.9, the maximum difference in power loss was found to be approximately 50%. Referring to figure 6, the film extent plots corresponding to load angles of 90° and 270° contrast the load bearing areas. High shear rates will occur in the full film region closer to the minimum film thickness. This contributes to the major part of the bearing power loss. Although shear power loss does occur in the cavitation region, its contribution to power loss is small when compared to the load carrying film region. This was due to the relatively high film thickness (and hence the shear rate) and the presence of air cavities.

For a constant eccentricity ratio, when the load is

acting in different directions, a large variation in oil flow rate is predicted. For a partially grooved bearing, the magnitude of oil flow rate is a combined effect of the feed pressure flow and the velocity induced flow. The observed general trend is that when the load is directed towards the grooved region, the oil flow rate is reduced considerably. This is due to :

(a) The axial oil flow rate is restricted due to the reduction of film thickness at the groove region.

(b) The film thickness at the groove end had been reduced which results in less lubricant being entrained into the bearing.

(c) Lubricant is forced back into the groove as the result of the generated hydrodynamic pressure close to the groove.

6 Effect of groove extent on steadily loaded journal bearing performance

For a fixed groove width to bearing width ratio ($a/b = 0.2$), four different groove extents were examined. These four groove extents were 60°, 120°, 180° and 360°. All grooves were located symmetrically above the vertical centre line and the load was acting vertically downwards.

Figure 7 shows that there is a marked difference in the load carrying capacity between the three partially grooved bearings and the fully grooved bearing. When the dimensionless supply pressure is gradually increased, the load carrying capacity is also increased and soon reaches a limiting value. No further effect on the fully grooved bearing load capacity is found when the supply pressure is increased further. For the other three partially grooved bearings, it is noted that the increase of supply pressure has a deleterious effect on the load capacity. This is because the hydrostatic load associated with the groove becomes more significant at high supply pressure. However, this effect is found to be smaller when the groove extent is smaller.

There is a negligible difference in power loss for the three partially grooved bearings concerned. As discussed previously, sufficient oil can be entrained into the bearing to form the load carrying film. The major part of the power loss is contributed by the high shear rate region in the load carrying film. The difference in power loss between the partially grooved bearings and the fully grooved bearing is due to the different bearing areas available to form the load carrying film.

Many researchers have found that in the case of a full central circumferential groove bearing, the oil flow rate is directly proportional to the supply pressure. The analysis of the three partially grooved bearings also showed a linear relationship (see figure 7), except at low dimensionless supply pressures. This observed characteristic seems to support the approach of superimposing the zero speed feed pressure flow (Q_p) on the zero feed pressure velocity induced flow (Q_v) to obtain the total bearing flow rate, a technique widely

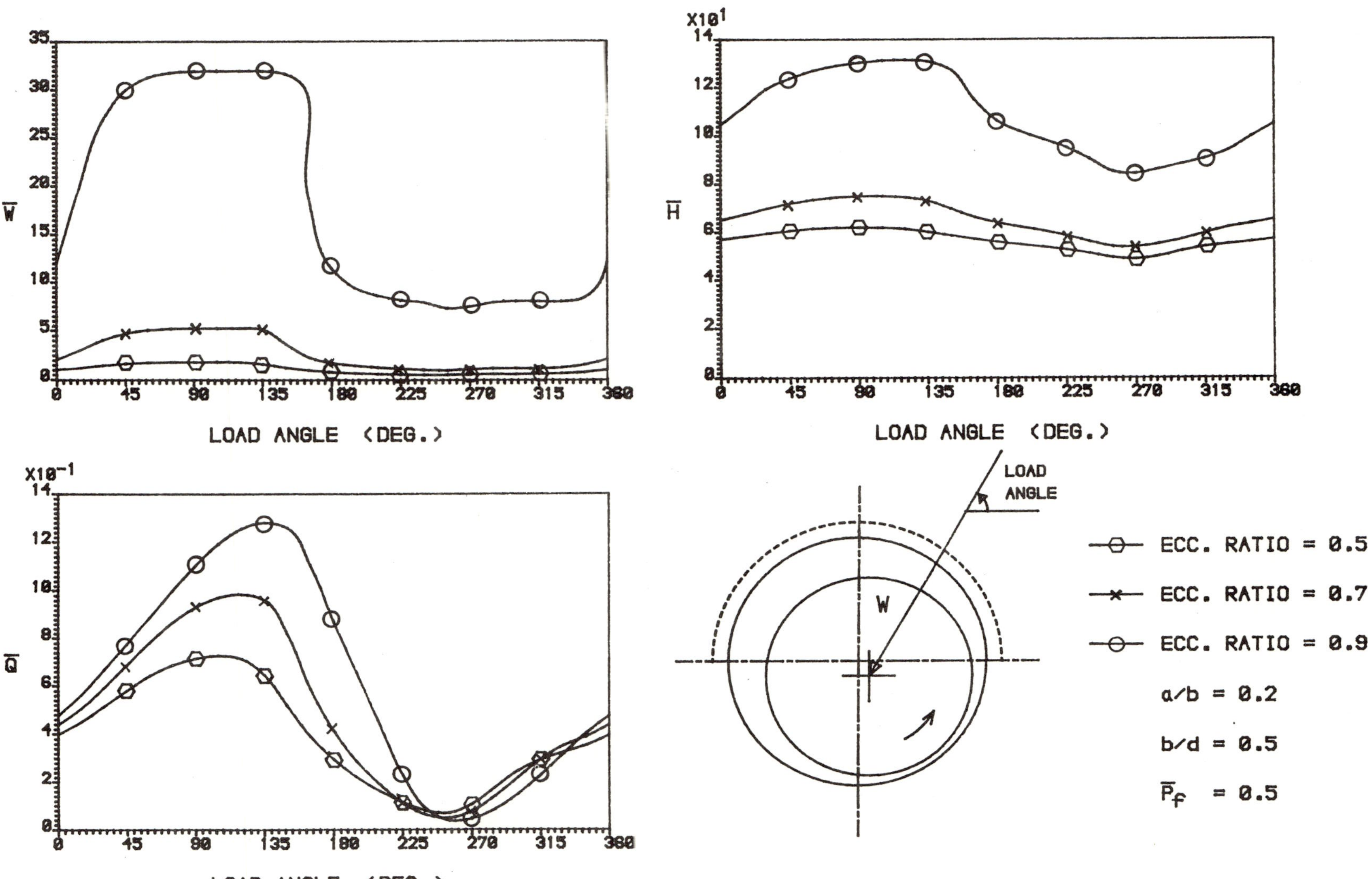

FIGURE (5)   INFLUENCE OF LOAD DIRECTION ON THE PERFORMANCE OF A 180 DEGREE
PARTIALLY GROOVED STEADILY LOADED JOURNAL BEARING

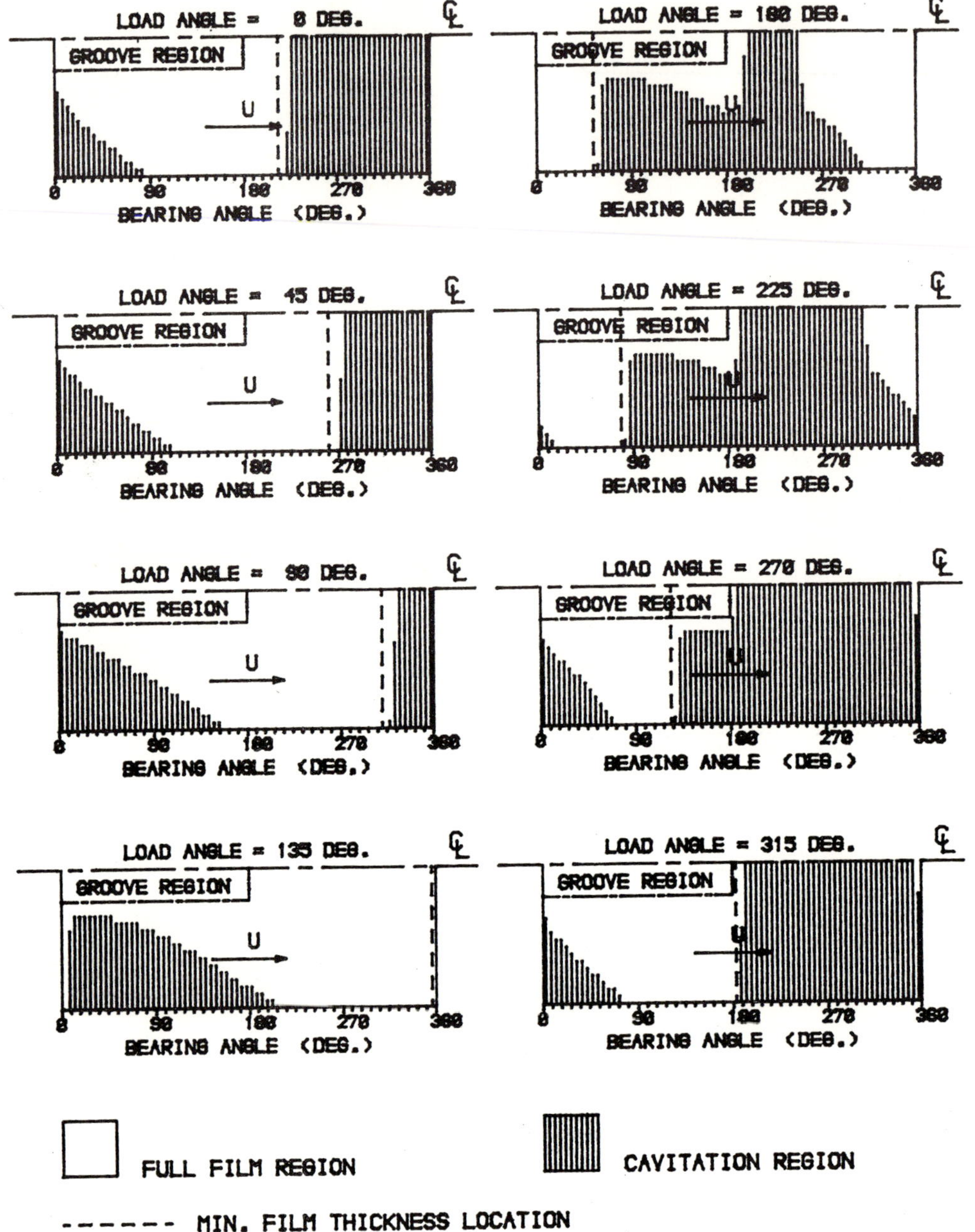

FIGURE (6) EFFECT OF LOAD DIRECTION ON FILM EXTENT FOR A 180 DEG.

PARTIALLY GROOVED STEADILY LOADED JOURNAL BEARING

$b/d = 0.5$, $a/b = 0.2$, $\overline{P}_f = 0.5$, ECC. RATIO = 0.7

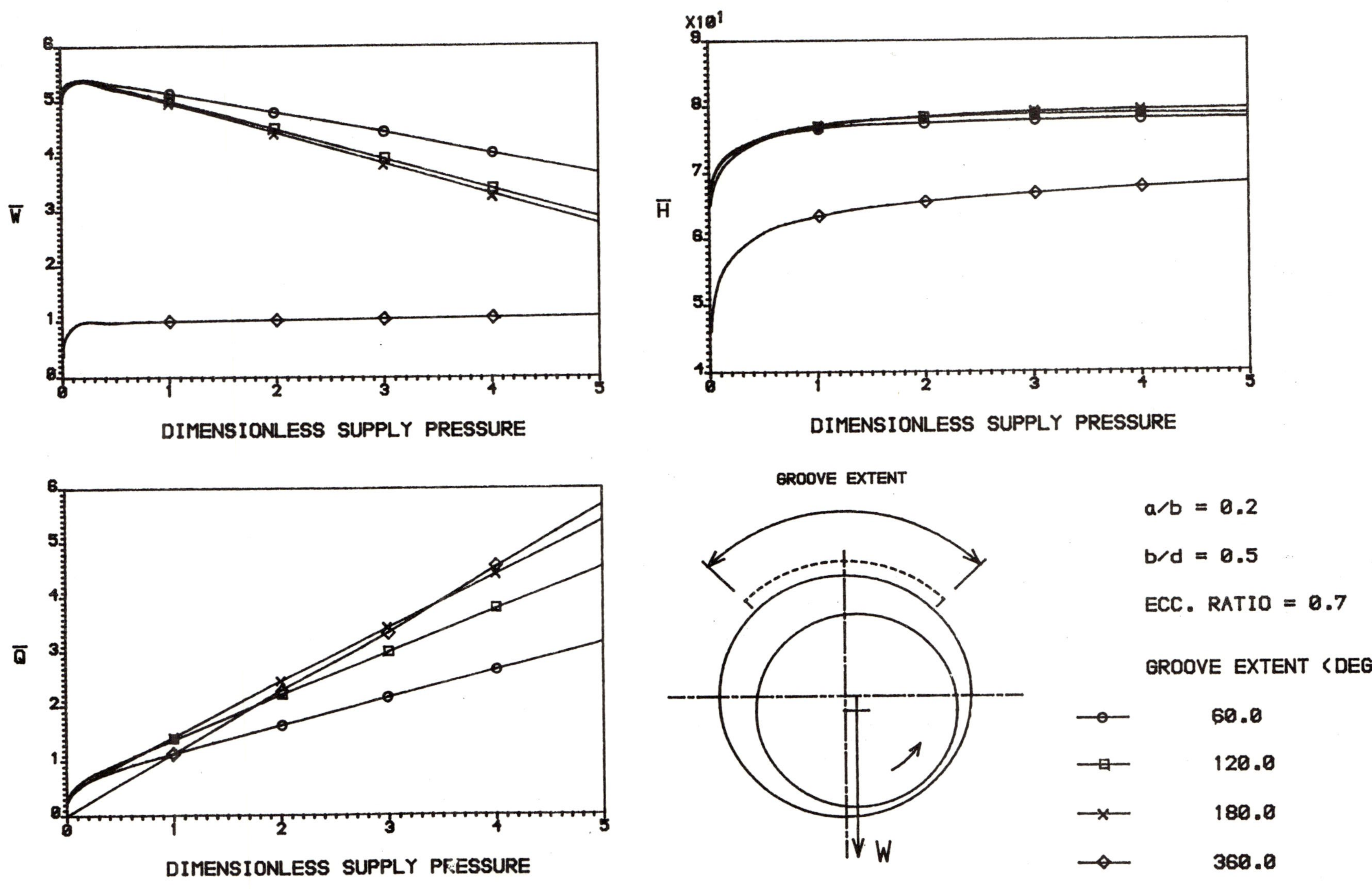

FIGURE (7) INFLUENCE OF GROOVE EXTENT ON BEARING PERFORMANCE

adopted in design procedures.

If there is no cavitation in the bearing, this approach will be precisely true, that is the total flow rate is given by;

$$Q_s = Q_v + Q_p \qquad (6)$$

Despite this observation, cavitation does occur in most bearings under normal operating conditions. Because of its simplicity and the absence of anything demonstrably better, this approach has been used for many years.

The validity and accuracy of this simple approach was studied in detail by Wilcock [9]. He reported that separately calculating and adding the two basic flow components could result in an overestimation of up to 20% in the bearing total flow rate. It was suggested that this discrepancy was due to the present of cavitation. However, the author's computed flow data in figure 7 have shown the existence of a linear relationship between the bearing side leakage flow and the feed pressure. To understand fully this linear behaviour, a further study on partial groove bearing flow rate will be required, and this is beyond the scope of the present study. It must be emphasized that the exact flow mechanism is a combined effect of feed pressure and hydrodynamic action. The superposition of the two basic flow components to form the total flow rate cannot be justified on a theoretical basis if cavitation occurs. The approach of simply adding the two basic flow components to give the total flow rate is only a simple convenience.

7 Conclusions

An analysis of partial groove steadily loaded journal bearings has been undertaken with a proper consideration of the cavitation boundaries. The main points can be summarized as follows;

- The load capacity of a partial groove bearing can be affected by high supply pressure.

- The oil entrainment at the partial groove end is an effective and reliable oil feeding mechanism.

- Dimensionless design data have been presented and further information can be found in reference [7].

- The influence of load direction has been examined. The bearing performance is highly affected by the journal position with respect to the partially grooved stationary bearing, and this in turn depends on the load direction.

- The presence of oil groove in the bearing load carrying region can seriously affect the load capacity.

- The influence of groove extent on bearing performance has been studied. The computed oil flow data of partial groove bearings shows a linear relationship with the supply pressure, except at the low range of supply pressure. This seems to support the common approach of superimposing the zero speed feed pressure flow on the zero feed pressure velocity induced flow to obtain the total bearing flow rate.

There are numerous specifications for partial groove bearings and it has only been possible to consider a narrow range. The data presented has been designed to give some physical insight into the performance of partial groove steadily loaded journal bearings. An important improvement for the future will be to developed a satisfactory approach to the thermal characteristics and to validate the design data experimentally.

Acknowledgment

The research described in this paper was part of a more comprehensive study of partial groove dynamically loaded journal bearing supported by Jaguar Cars Ltd. The authors would like to acknowledge the kind permission of the company to publish this paper and to thank Mr. S H Richardson, Mr. M Roskilly and Mr. R Lee for their support.

References

[1] Calculation methods for steadily loaded axial groove hydrodynamic journal bearings. ESDU International plc, Item Number 84031, 1984.

[2] Calculation methods for steadily loaded central circumferential groove hydrodynamic journal bearings. ESDU International plc, Item Number 90027, 1990.

[3] **Dowson, D.** and **Taylor, C.M.** Cavitation in bearings. Annual Review of Fluid Mechanics, Vol. 11, 1979. pp 35-66.

[4] **Jakobsson, B.** and **Floberg, L.** The finite journal bearing considering vaporization. Trans. Chalmers Univ. of Tech., 1957, pp 190.

[5] **Rowe, W.E.** and **Chong, F.S.** A computer algorithm for cavitating bearings. Trib. Intl., Vol. 17, No. 5, 1984, pp 243-250.

[6] **Patankar, S.V.** Numerical heat transfer and fluid flow. McGraw Hill, New York. pp 52-54.

[7] **Lai, C.K.K.** A study of partial central circumferential groove engine bearings. Ph.D. thesis, Dept. of Mech. Eng., Univ. of Leeds, in preparation.

[8] **Clayton, G.A.** and **Taylor, C.M.** Design data for the steadily loaded full central circumferential grooved plain journal bearing. Proc. Instn. Mech. Engrs., Vol. 204, 1990, pp 53-61.

[9] **Wilcock, D.F.** Influence of feed groove pressure and related lubricant flow on journal bearing performance. STLE, Trib. Trans., 1988, Vol. 31, 3, 398-404.

A theoretical and experimental study of thermal effects in a plain circular steadily loaded journal bearing

M T MA, BEng, MEng and C M TAYLOR, BSc, MSc, PhD, DEng, CEng, FIMechE
Institute of Tribology, Department of Mechanical Engineering, University of Leeds, UK

SYNOPSIS *Accurate predictions of the temperature profile in a steadily loaded journal bearing are not straightforward. In the present study, a rigorous numerical thermal analysis is developed to enable the prediction of detailed temperature mapping. The model includes a particular consideration of the mixing of the recirculating lubricant with the incoming supply in the vicinity of grooves. A 110 mm diameter cylindrical bearing having two axial grooves has been tested at specific loads up to 4 MPa and rotational frequencies up to 120 Hz. Power loss and flow rate were measured directly and automated data aquisition of temperature information has been undertaken. Some of the experimental data will be presented and compared with the predictions of the current model. In addition, the predictions of the current model are checked with those of a published design procedure [1].*

NOTATION

A_v, B_v	oil dependent constants
C	radial clearance
C_g	empirical coefficient (see Equation (14))
C_p	lubricant specific heat
d	journal diameter
h	local film thickness
K_0	lubricant thermal conductivity
K_B	bush thermal conductivity
L	bearing length
L_g	axial groove length
N	rotational frequency
p	local film pressure
p_f	oil feed pressure
p_s	specific load
Q	volume flow rate
R	journal radius
R_2	outer radius of bush
T	local film temperature
Tb	local bush temperature
T_a	ambient temperature
T_s	shaft surface temperature
U	journal surface velocity
u, v	velocity components in x,y direction respectively
W	applied or external load
x, y, z	Cartesian coordinates
θ, r, z	polar coordinates
θ	circumferential coordinate measured from middle of upper lobe
ρ	lubricant mass density
ν	lubricant kinematic viscosity
η	lubricant dynamic viscosity
ϕ	angular extent of groove
ψ	clearance ratio
λ_g	groove wall heat transfer coefficient
λ_o	outer bush surface heat transfer coefficient

Subscripts

in	inlet
c	cavitation
f	feed
g	groove
p	lobe or groove number
r	reverse flow
t	trailing edge of lobe

1   INTRODUCTION

Traditional journal bearing design is usually based upon isothermal lubrication theory. This is not adequate for many situations. A considerable amount of heat will be generated in the film at high rotational speeds due to viscous dissipation and consequently a large temperature variation will be created in the bearing. This can result in a potential decrease in load-carrying capacity of the bearing. Also, the bearing bush is usually lined with a material which cannot satisfactorily sustain a high temperature. For example, whitemetal can only tolerate a maximum temperature of about 120 $^\circ C$ [1]. Moreover, if the maximum lubricant temperature exceeds a critical value, excessive oxidization may occur.

Hence the maximum temperature itself has been considered as one of the main bearing design criteria and industry is currently demanding more accurate assessment of bearing material temperatures as critical operating conditions are approached. Therefore, in order to investigate the thermal behaviour of journal bearings both thermohydrodynamic analyses and experiments have been considerably developed during the last few decades. A comprehensive literature review up to the mid-eighties can be found in [2].

A review of the published literature reveals that the thermohydrodynamic analyses have been established with the methods varying from approximate approaches [3-9] to full numerical solutions [10-16]. However, there still remain special complexities which have not yet been resolved adequately, such as oil groove mixing, thermal behaviour of the fluid in cavitation regions, temperature boundary conditions and system interactions. These can cause difficulties in developing a thermal analysis. Concerning thermal effects in plain journal bearings, much experimental work has been undertaken by a number of researchers [9,10,12,17-20]. The parameters which have been measured included bush and shaft surface temperatures, power loss, flow rate, fluid film pressure and film thickness.

The general objectives of the present study were: (1) To develop a detailed thermal analysis of journal bearings with a particular consideration of oil groove mixing; (2) To provide some experimental information of the thermal behaviour of a two-axial-groove cylindrical bore bearing; (3) To verify and validate the current numerical model through comparison with the experimental data. In addition, the predictions of the current model have been compared with those of ESDU Item 84031 [1]. The software used in this comparison did not incorporate the amendment A to this item introduced in 1991.

2 THEORETICAL FORMULATION

2.1 Governing Equations

The governing equations are subject to the usual assumptions. A particular assumption based upon experimental evidence is that there is no axial temperature variation in both the film and bush. This can greatly simplify the analysis.

To obtain the film pressure field Reynolds equation has to be solved. The general form of the Reynolds equation can be written as,

$$\frac{\partial}{\partial x}\left(F_2 \frac{\partial p}{\partial x}\right) + \frac{\partial}{\partial z}\left(F_2 \frac{\partial p}{\partial z}\right) = U \frac{\partial}{\partial x}\left(h - \frac{F_1}{F_0}\right) \qquad (1)$$

where,

$$F_0 = \int_0^h \frac{dy}{\eta}$$

$$F_1 = \int_0^h \frac{y}{\eta} dy$$

$$F_2 = \int_0^h \frac{y}{\eta}\left(y - \frac{F_1}{F_0}\right) dy$$

To achieve the temperature field in the film, the following simplified energy equation is solved on the centreplane of the bearing.

$$\rho C_p \left(u \frac{\partial T}{\partial x} + v \frac{\partial T}{\partial y}\right) = K_0 \frac{\partial^2 T}{\partial y^2} + \eta \left[\left(\frac{\partial u}{\partial y}\right)^2\right] \qquad (2)$$

If the heat conduction in the axial direction is ignored, the heat conduction equation in the bush can be written as

$$r^2 \frac{\partial^2 Tb}{\partial r^2} + r \frac{\partial Tb}{\partial r} + \frac{\partial^2 Tb}{\partial \theta^2} = 0 \qquad (3)$$

The relationship between kinematic viscosity and temperature can be determined by the Walther equation with the kinematic viscosity in (cS) and the temperature in Celsius, that is,

$$\lg\left[\lg\left(\nu + 0.8\right)\right] = A_v + B_v \lg(T + 273) \qquad (4)$$

where, A_v and B_v are oil dependent constants.

2.2 Boundary Conditions

The above equations are subject to a set of boundary conditions, which can be described as follows:

2.2.1 Pressure Boundary Conditions

In the fluid film the classical Reynolds boundary conditions are applied, i.e.,

$$p|_{x=x_{c,p}} = 0 , \quad \text{and} \quad \frac{\partial p}{\partial x}\Big|_{x=x_{c,p}} = 0 \qquad (5)$$

At the inlet and outlet of a lobe the pressures are equal to the oil feed pressure over the full bearing width,

$$p|_{x=x_{in,p}} = p_f , \quad p|_{x=x_{t,p}} = p_f \qquad (6)$$

At the two sides,

$$p|_{z=\pm\frac{L}{2}} = 0 \qquad (7)$$

2.2.2 Temperature Boundary Conditions

On the interface between the fluid film and the bush the heat flux continuity condition is used,

$$K_B \frac{\partial Tb}{\partial r}\Big|_{r=R} = -K_0 \frac{\partial T}{\partial y}\Big|_{y=0} ,$$
$$\text{and} \quad T|_{y=0} \equiv Tb|_{r=R} \tag{8}$$

On the interface between the shaft and the fluid film the temperature is assumed to be constant and the no-net-heat flow condition is applied,

$$\int_0^{2\pi R} K_0 \frac{\partial T}{\partial y} dx = 0 ,$$
$$\text{and} \quad T|_{y=h} \equiv T_s = constant \tag{9}$$

The oil temperature at the inlet is considered to be constant and determined by a new oil groove mixing model, which will be presented in the next subsection.

The free convection hypothesis (i.e., natural heat transfer without forced cooling) is applied to the outside surface of the bearing and its related interfaces in the groove regions. The outside surface of the bearing is exposed to ambient conditions, thus the temperature on this surface can be determined by the following expression:

$$K_B \frac{\partial Tb}{\partial r}\Big|_{r=R_2} = \lambda_o \left(T_a - Tb|_{r=R_2} \right) \tag{10}$$

Two walls adjacent to a groove are exposed to the fresh feed oil in the groove. The temperatures on the walls are calculated respectively from the following equations:

On the inlet wall of lobe p,

$$K_B \frac{\partial Tb}{\partial x}\Big|_{x=x_{in,p}} = \lambda_g \left(Tb|_{x=x_{in,p}} - T_{f,p} \right) \tag{11}$$

On the outlet wall of lobe p,

$$K_B \frac{\partial Tb}{\partial x}\Big|_{x=x_{t,p}} = \lambda_g \left(T_{f,p+1} - Tb|_{x=x_{t,p}} \right) \tag{12}$$

2.3 Oil Groove Mixing Model

As a boundary condition the inlet temperature plays an important role in determining the thermal behaviour of hydrodynamic bearings. However, this is not easy to establish accurately since there exists a complex hot and cold oil mixing in the vicinity of an oil groove. Therefore the problem is still not adequately resolved though some valuable work has been done by several researchers [12, 21-23]. In this subsection, a robust oil groove mixing model is developed by applying the energy balance approach and the principles of energy and mass conservation to evaluate the inlet film temperature for multi-lobed journal bearings.

2.3.1 Energy Conservation Equation

Basically, lubricant flow in an oil groove consists of four components comprising the flow leaving the trailing edge of the upstream lobe ($Q_{t,p-1}$), the flow being fed into the groove ($Q_{f,p}$), the flow entering the downstream lobe ($Q_{in,p}$) and the flow leaking from the ends of the groove ($Q_{g,p}$) (see Appendix 1). However, when the bearing operates at a large eccentricity reverse flow may occur in the vicinity of the inlet of a lobe due to the significant pressure gradients. A certain amount of lubricant ($Q_{r,p}$) which enters the lobe is subsequently reversed and discarded as Figure 1(a) shows. When this amount constitutes a measurable fraction of the inflow, it should be taken into account in the mixing process. From the viewpoint of generality, the effect of reverse flow will be included in the establishment of the current oil groove mixing model.

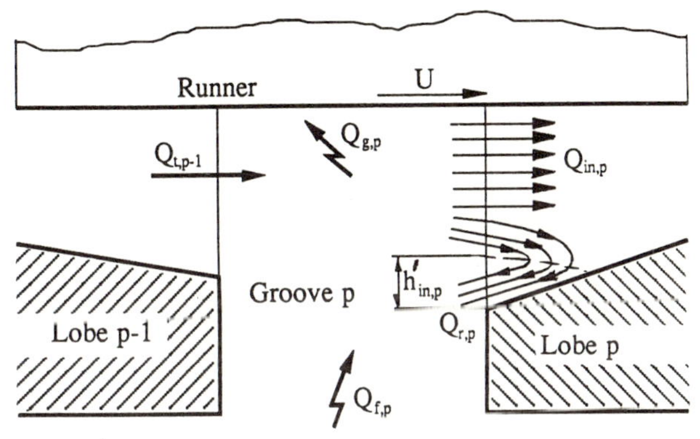

(a) Side view of oil groove

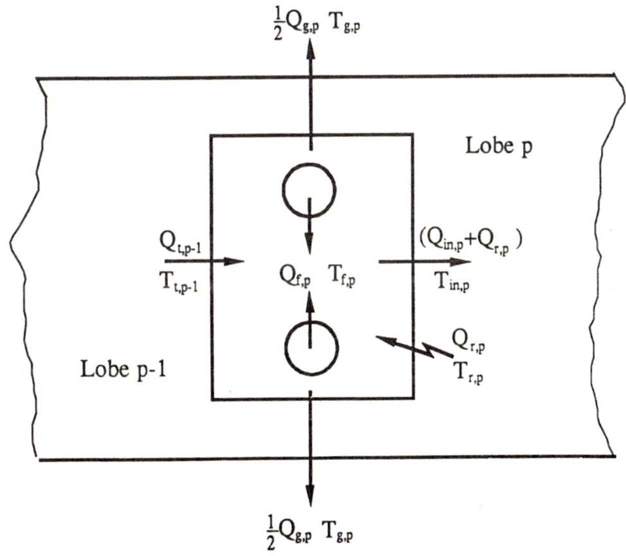

(b) Top view of oil groove

Figure 1 Oil groove mixing including reverse flow

It is assumed that each component of oil flow is at a constant temperature. From Figure 1(b) the heat balance equation in an oil groove can be obtained on the basis of the law of energy conservation,

$$\rho C_p Q_{t,p-1} T_{t,p-1} + \rho C_p Q_{f,p} T_{f,p} + \rho C_p Q_{r,p} T_{r,p}$$
$$= \rho C_p (Q_{in,p} + Q_{r,p}) T_{in,p} + \rho C_p Q_{g,p} T_{g,p} \tag{13}$$

where, $T_{r,p}$ is average lubricant temperature of the reverse flow.

Assuming

$$T_{g,p} = T_{f,p} + C_g(T_{in,p} - T_{f,p}) \tag{14}$$

and noting that the oil density (ρ) and specific heat (C_p) can be take as constant, then we have

$$T_{in,p} =$$
$$\frac{Q_{t,p-1} T_{t,p-1} + Q_{r,p} T_{r,p} + [Q_{f,p} - (1 - C_g) Q_{g,p}] T_{f,p}}{Q_{in,p} + Q_{r,p} + C_g Q_{g,p}} \tag{15}$$

where, C_g is an empirical coefficient from which the heat transfer effect in the groove region may be included, and $0 \leq C_g \leq 1.0$.

2.3.2   Mass Conservation Equation

In a groove the oil flows should be balanced, so the following mass conservation equation exists no matter whether reverse flow takes place or not,

$$\rho Q_{t,p-1} + \rho Q_{f,p} = \rho Q_{in,p} + \rho Q_{g,p} \tag{16}$$

where, $Q_{t,p-1} = Q_{c,p-1}$.

Hence the feed oil flow rate can be calculated by,

$$Q_{f,p} = Q_{in,p} + Q_{g,p} - Q_{t,p-1} \tag{17}$$

2.3.3   Brief Discussion

If $Q_{f,p} \leq 0$, a case which could occur in multi-lobed journal bearings, the groove is considered to be a sink and the inlet film temperature of lobe p is equal to the outlet film temperature of lobe p-1, that is,

$$T_{in,p} = T_{t,p-1}$$

Under this circumstance,

$$C_g = 1 \quad \text{and} \quad T_{g,p} = T_{in,p}$$

If $Q_{t,p-1} = 0$, (this is of no practical significance but of theoretical interest), there will be no oil mixing, then,

$$C_g = 0$$

and for no reverse flow cases,

$$T_{in,p} = T_{g,p} = T_{f,p}$$

In the present work, $C_g = 0.5$ has been used for the calculations and has been shown to be appropriate.

2.4   Global Solution Procedure

The governing equations were normalized and discretized using the finite difference method. In order to handle the reverse flow situations, the energy equation and the heat conduction equation were solved separately by imposing the heat flux continuity condition at the interface. The global solution procedure is as follows:

1. For the prescribed operating conditions, an initial eccentricity ratio, attitude angle, pressure and temperature field are assumed.

2. The Reynolds equation is solved iteratively using the successive-over-relaxation technique. During the computations, all negative gauge pressures are set equal to zero to implement the Reynolds boundary conditions. Having achieved the pressure field, the attitude angle is checked and, if necessary, modified and then the procedure is iterated until the relative difference of the vertical resultant force component between two successive iterations is less than 0.1 percent.

3. With the new pressure values, the energy equation is solved by the so-called *Sweep Method* (i.e., a modified Gaussian elimination technique), including the special treatment of the reverse flow situations. After convergence of the film temperature field, the heat conduction equation is solved also using the sweep method. As soon as the whole temperature field has been obtained, the heat flux continuity condition at the film-bush interface is checked. If this condition is not satisfied, the computation for the film and bush temperatures will be repeated until the relative error of each local interface temperature between two successive iterations is below 0.1 percent. Then the shaft and mixing inlet temperatures are calculated and with these new values the solution procedure for the temperature field will be repeated until the convergence is obtained.

4. With the current temperature values, the pressure field is updated and the entire cycle is repeated until both the pressure and temperature fields converge.

3 EXPERIMENTAL WORK

3.1 General Description of the Apparatus

An experimental apparatus has been developed to test 110 diameter plain journal bearings. Figure 2 shows a schematic arrangement of the apparatus. The test shaft was supported by two oil lubricated rolling ball bearings. The shaft was driven by a 11.2 kW (15 HP) d.c. motor through an infinitely variable transmission and a toothed drive belt. The rotational frequency could reach 120 Hz.

The load was applied to the test bearing by four hydraulic cylinders underneath the housing. Each cylinder had an effective area of 640 mm^2. A hand operated pump was used to inject the hydraulic fluid into the cylinders. Between the housing and cylinders, four linear roller bearing assemblies running between hardened races were fitted to allow a horizontal freedom.

Two separate oil supply systems were adopted to feed the oil into the test bearing and support bearings respectively. A variable delivery oil pump was used in each supply system. The oil supply temperature to the test bearing was maintained by a thermostat and an oil cooler working with the recirculating water. The oil supply pressure to the test bearing was also maintained by adjusting the controller to the pump through a pressure gauge fitted in the supply pipe.

In order to account for the friction in the support bearings, the test bearing was replaced by two identical rolling element ball bearings of the same type and size as those used to react the applied load, with the adoption of an alternative shaft arrangement. In this manner the measured friction could be equally apportioned between the four bearings used in calibration and the contribution of the support bearings as a function of load, speed and temperature identified.

3.2 Instrumentation

The applied load was measured by monitoring the hydraulic fluid pressure to the cylinders with a pressure gauge. The rotational speed and torque were measured by transducers. The transducers were inserted into the drive shaft using a pair of couplings. The readings of the rotational speed, torque and power loss could be directly obtained from the associated electronic indicator. The lubricant flow was measured by putting the drain pipe into a bucket and a stop watch was to determine the rate.

The temperatures were measured by Iron-Constantan thermocouples. Those located in the whitemetal were 0.5 to 1.0 mm from the lubricant interface. Figure 3 illustrates the thermocouple locations in the bush and grooves. The detailed temperature information was acquired by ADCs on a VAX 8600 computer through a multiplexer unit.

3.3 Experimental Procedure

A two-axial-groove cylindrical bore bearing has been tested. The bearing dimensions and test conditions are listed in Tables 1 and 2 respectively. The results were recorded after thermal equilibrium was established. All the relevant data were written into a named data file on the computer.

Table 1: **Test bearing dimensions**

Nominal journal diameter :	$d = 110$ mm
Aspect ratio :	$L/d = 0.7$
Clearance ratio :	$\psi = 0.0017$
Angular groove extent :	$\phi = 55$ degrees
Outer radius of bush :	$R_2 = 100$ mm
Axial groove length :	$L_g = 60$ mm

Table 2: **Test conditions**

Oil feed pressures :	$67\ kN/m^2$ (10 psi), $134\ kN/m^2$
Oil feed temperature :	$45\ ^\circ C$
Lubricant :	ISO VG 32
Specific loads :	0.2, 1.0, 2.0, 3.0, 4.0 MPa
Shaft rotational frequencies :	1000, 2500, 4000, 5500, 7000 rpm

4 RESULTS AND DISCUSSION

In this section, some of the experimental results are presented and compared with the predictions of the current model. The thermophysical properties adopted for the calculations are given in Table 3. In addition, the predictions of the current model are also checked with those of [1].

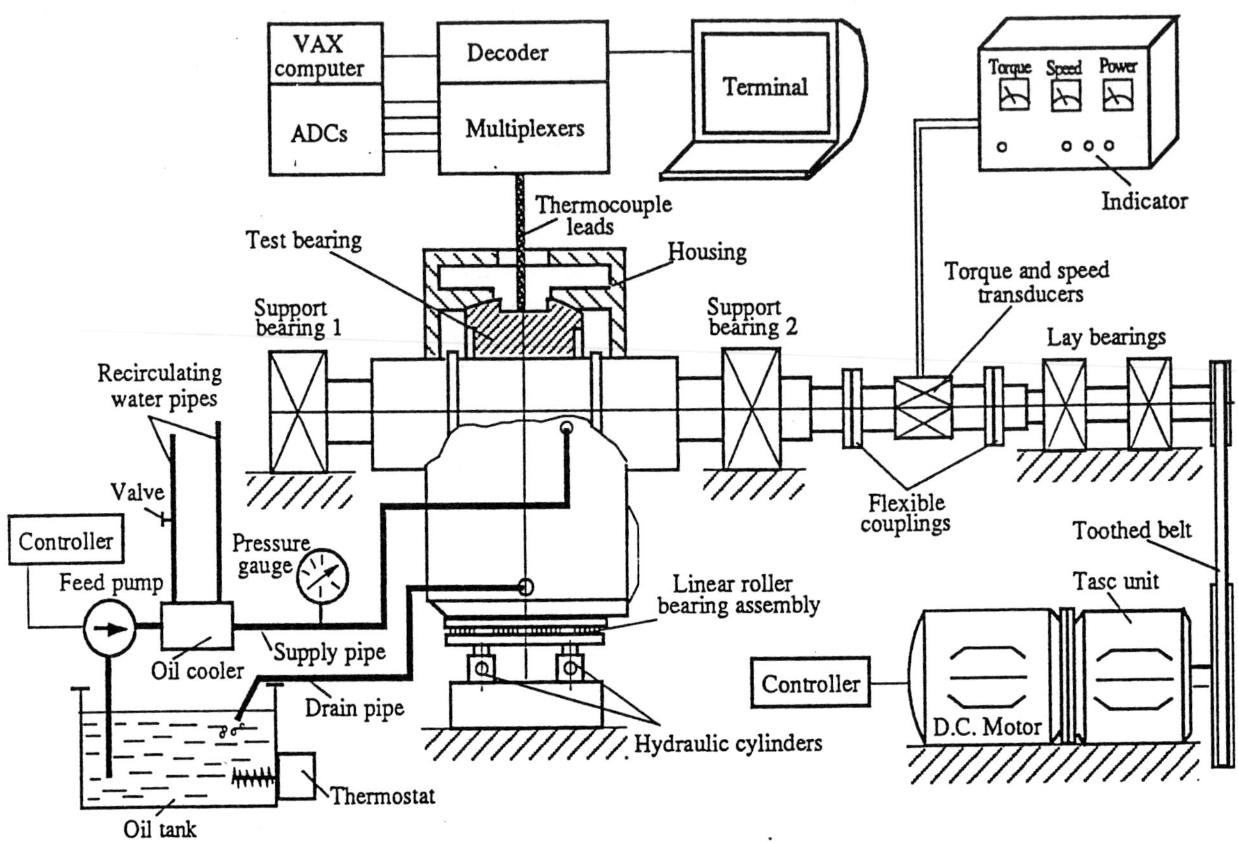

Figure 2 Schematic arrangement of the experimental apparatus

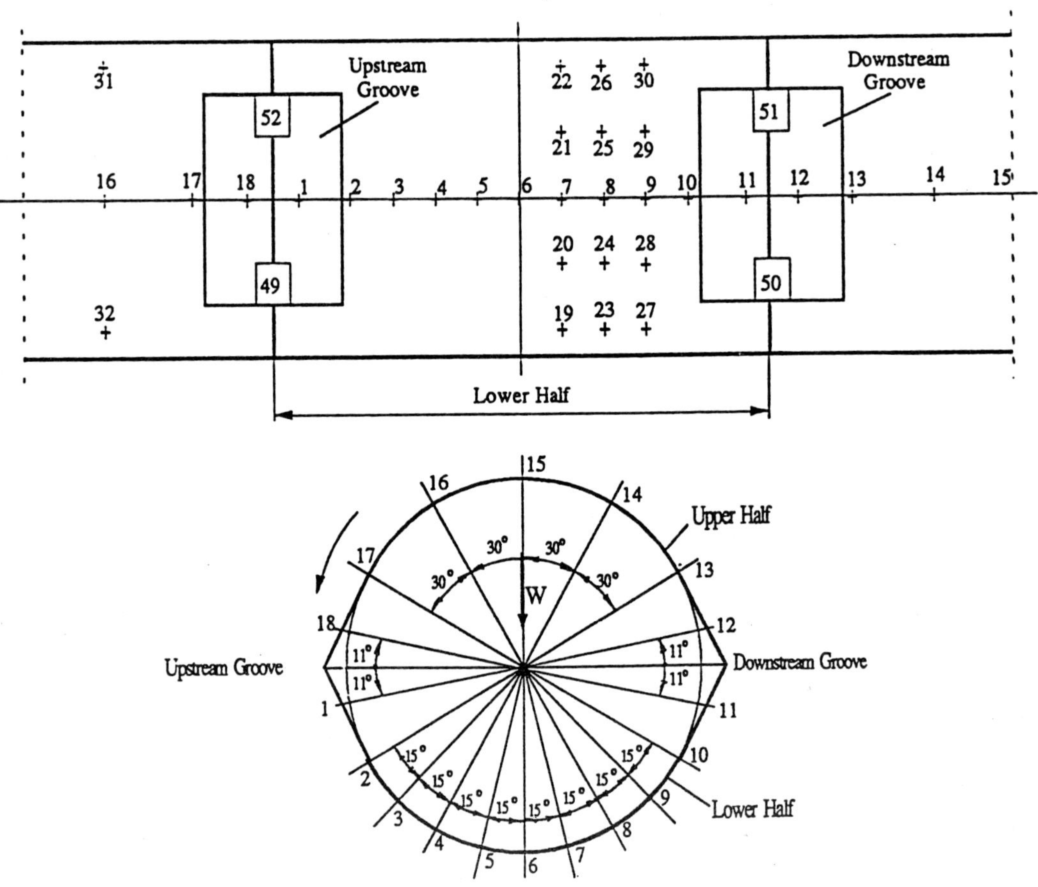

Figure 3 Thermocouple locations in the bushes and grooves

Table 3: **Thermophysical properties adopted**

$K_0 = 0.13\ (W/m^oC)$	$K_B = 52\ (W/m^oC)$
$\lambda_g = 100\ (W/m^{2o}C)$	$\lambda_o = 55\ (W/m^{2o}C)$
$\rho = 850\ (kg/m^3)$	$C_p = 2000\ (J/kg^oC)$
$A_v = 9.455$	$B_v = -3.176$
$T_a = 45\ (^oC)$	

4.1 Detailed Thermal Behaviour

Figure 4 (a) and (b) shows the comparison of the bush surface temperature excursions on the bearing centre-plane in the two lobes for a light load of 1.694 kN (p_s=0.2 MPa) and at the operating speeds of 2500 and 4000 rpm respectively. It is clear that the predicted and measured temperature profiles agree extremely well in both lobes. The temperature excursions are not significant and are similar in the two lobes. This is due to the comparatively uniform film and the dominant effect of hot oil carryover at the light load. Moreover, the predicted and measured temperatures at the inlets are very similar and noticeably higher than the feed temperature. This may imply two points: firstly that the effect of hot oil carryover or oil groove mixing is important in the determination of the temperatures; the secondly that the current oil groove mixing model is adequate as the correlation is so close between the theory and experiment.

Figure 5 (a) and (b) illustrates the comparison of the predicted and measured centre-plane temperatures for a moderate load of 16.94 kN (p_s=2 MPa) and with rotational speeds of 2500 and 4000 rpm respectively. It can be clearly seen that the correlation between the theoretical and experimental results is very close in both the loaded and unloaded lobes. Under this heavier loading condition the temperature excursions in the two lobes are greatly different and generally the temperatures in the loaded lobe are much higher than those in the unloaded lobe. In the loaded lobe the changes in temperature are significant from the inlet to the outlet, while in the unloaded lobe, the temperature rise is modest. This is an expected result, since pressurisation exists and the film is thinner in the loaded lobe, thus the viscous dissipation in this lobe is much more significant than that in the unloaded lobe which has a divergent film wedge. In comparing graphs (a) and (b), it can be seen that the temperatures and thermal excursions are larger for the higher speed of 4000 rpm.

Figure 6 (a) and (b) compares the temperature distributions for a very heavy load of 33.88 kN ($p_s = 4$ MPa) and at the two different speeds. Once again, the good agreement can be noted between the theory and experiment over the whole bearing circumference. It is notable that the temperatures at the inlet of the unloaded lobe are very close to the feed value. This is

due to the fact that at the heavy load the bearing operated with a large eccentricity and consequently, only a small amount of hot oil was carried over from the loaded lobe. Thus the cold supply oil has dominated in the determination of the effective inlet film temperature for the unloaded lobe. Furthermore, it can be clearly seen that there is a considerable drop in temperature in the outlet region of the loaded lobe. This is known as the temperature fade [24] and even more significant with non-circular bearings [25]. In conventional thermal analyses, no such temperature fade has been predicted although much evidence has been shown in some experimental tests [9,17]. However, it is clear that the current detailed thermal analysis can effectively predict this temperature fade. The details concerning this will be addressed in a separate paper in due course, however, a summary is presented in Appendix 2.

4.2 Global Thermal Performance

In this subsection, the thermal performance characteristics of the bearing, including the maximum bush temperature, power loss and flow rate, are presented and the effects of specific load and rotational speed are examined. The experimental results are compared with the predictions of the current model.

Figure 7 shows the predicted and measured bearing static performance over a range of specific loads up to 4 MPa for a fixed rotational frequency of 4000 rpm and a feed pressure of 134 kPa. In graph (a), it can be seen that the predicted maximum bush temperatures agree very well with the measured data. Clearly the maximum bush temperature increases gradually with increase in load. The comparison of power loss is illustrated in graph (b) of the figure. The close agreement can be noted. The power loss increases smoothly with specific load.

The variation of the flow rate with specific load is interesting. The flow rate increases markedly in the lower range of specific loads, then almost levels off or decreases slightly. This can be explained as follows: In the lower range of loads, the increasing flow rate is due to the sharp increase in film temperature and axial pressure gradients at the ends of the bearing. For higher loads, although the film temperature and pressure gradients still increase with specific load, meanwhile the film extent reduces significantly, and the combined effect is that the total flow rate hardly changes in the higher load range. From graph (c), some discrepancies can be noted between the prediction and measurement with the predicted flow rates being moderately lower than the measured data. This may be due to the fact that the effect of film reformation was not taken into account in the estimation of the groove pressure flow rate (see Appendix 1). Film reformation in the vicinity of a groove is more likely to occur at such a high feed pressure and this will lead to a increase in

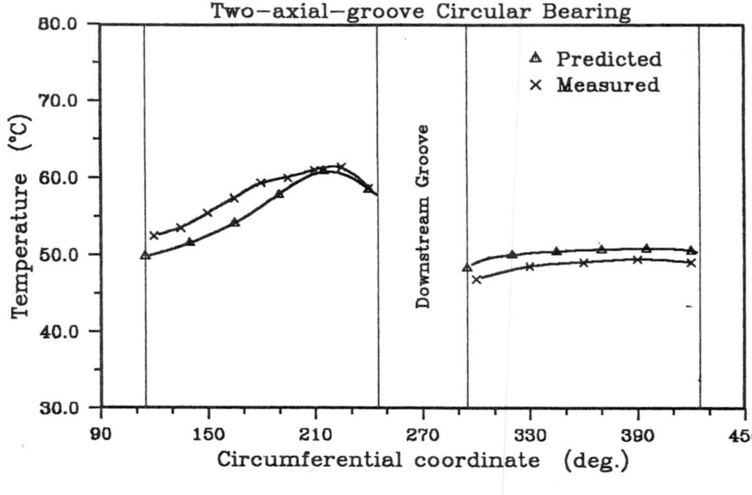

Figure 4 (a) Comparison of the predicted
and measured centre plane temperatures
W=1.694 (kN), N=2500 (rpm), p_r=134 (kPa), T_r=45 (°C)

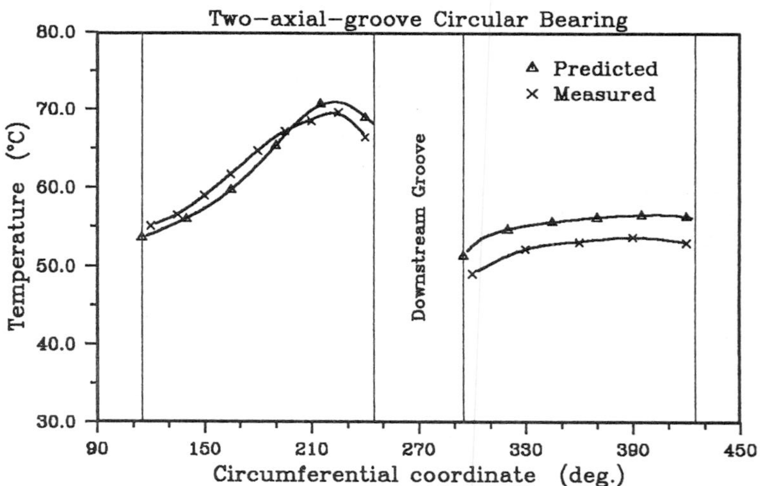

Figure 5 (a) Comparison of the predicted
and measured centre plane temperatures
W=16.94 (kN), N=2500 (rpm), p_r=134 (kPa), T_r=45 (°C)

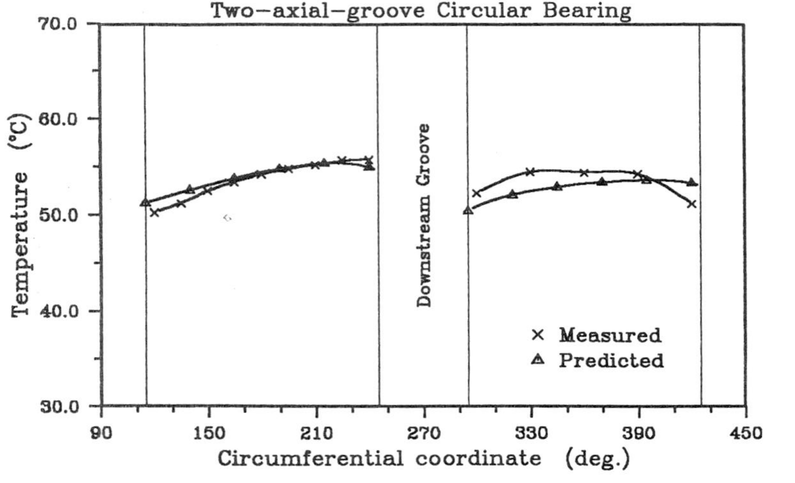

Figure 4 (b) Comparison of the predicted
and measured centre plane temperatures
W=1.694 (kN), N=4000 (rpm), p_r=134 (kPa), T_r=45 (°C)

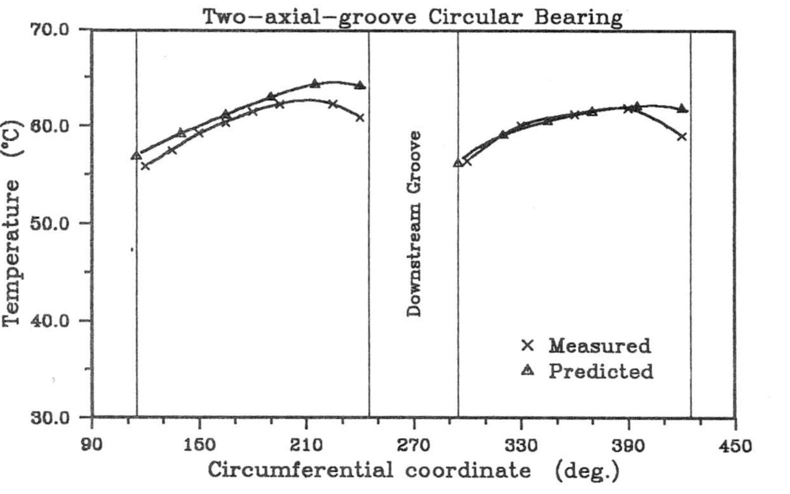

Figure 5 (b) Comparison of the predicted
and measured centre plane temperatures
W=16.94 (kN), N=4000 (rpm), p_r=134 (kPa), T_r=45 (°C)

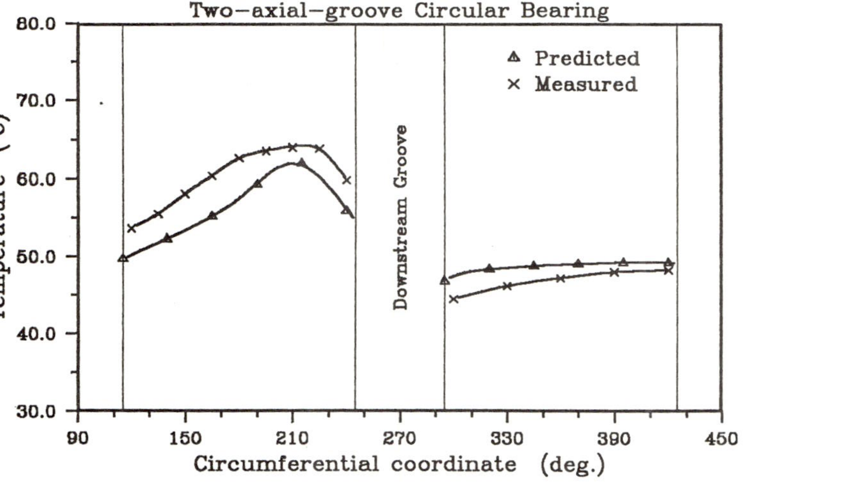

Figure 6 (a) Comparison of the predicted
and measured centre plane temperatures
W=33.88 (kN), N=2500 (rpm), p_f=134 (kPa), T_f=45 (°C)

Figure 6 (b) Comparison of the predicted
and measured centre plane temperatures
W=33.88 (kN), N=4000 (rpm), p_f=134 (kPa), T_f=45 (°C)

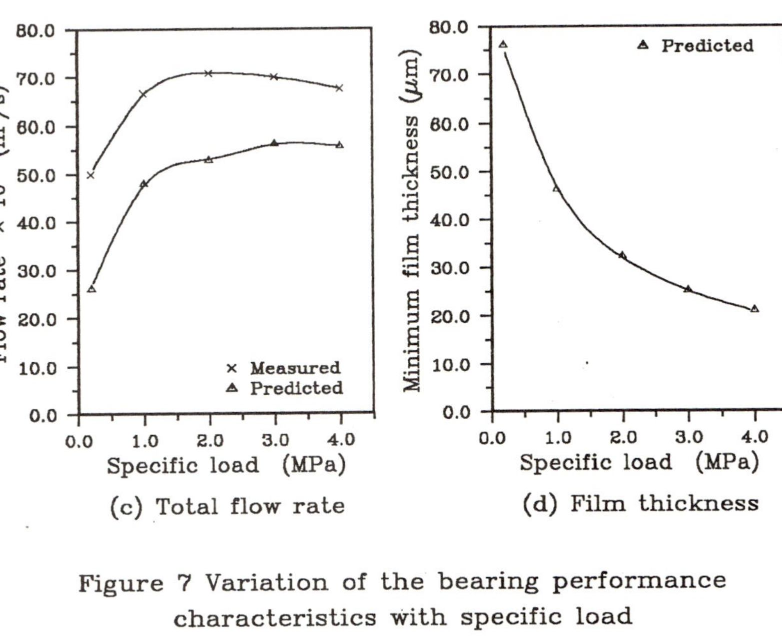

(a) Maximum temperature

(b) Power loss

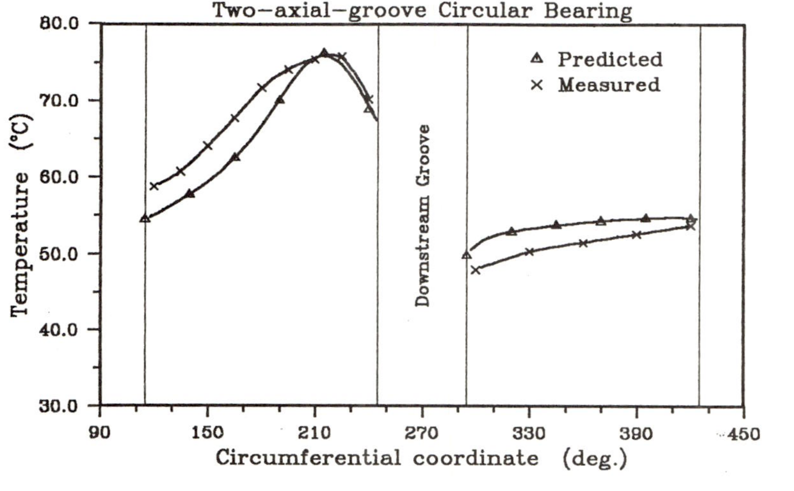

(c) Total flow rate

(d) Film thickness

Figure 7 Variation of the bearing performance
characteristics with specific load
N=4000 (rpm), ψ=0.0017, p_f=134 (kPa), T_f=45 (°C)

the pressure flow from the groove.

Since the minimum film thickness could not be measured, only the calculated results are presented in graph (d). Clearly the predicted minimum film thickness reduces significantly with increase in specific load.

In Figure 8, the bearing performance is presented for a feed pressure of 67 kPa, which is half of that for the results presented in Figure 7, in order that the effects of feed pressure might be examined. In comparing the corresponding graphs in Figures 7 and 8, except for the flow rate the others are very similar. This means that the maximum temperature, power loss and minimum film thickness are hardly affected by feed pressure. As shown in graph (c) of the figure, the agreement between the predicted and measured flow rates is surprisingly good. This indicates that the estimation of the groove pressure flow rate is accurate for this lower feed pressure.

Figure 9 shows the variation of the performance of the bearing with rotational speed for a fixed load of 16.94 kN. As graph (a) illustrates, generally the current model predicts the maximum bush temperature with fair accuracy. However, at the extreme speeds some discrepancies can be noted and the model overestimates the maximum temperature at high rotational frequencies. This may be principally due to the following two factors: (1) Thermal distortion of the bush: at the high speeds the temperature excursions were large in the bearing. In consequence, the thermal distortion of the bush might become significant. Hence the gap between the bush and shaft surfaces would be larger than the nominal one and this could substantially lower the operating temperatures of the bearing. (2) Cooling effect of the supply oil: since the cold supply lubricant was not directly fed into the grooves but passed through two channels on the outer surface of the bottom half of the bearing before reaching the grooves, there is no doubt that a certain amount of heat would be absorbed by the supply lubricant from the bearing. Hence the measured maximum bush temperature could be lower.

As might be expected, both the predicted and measured maximum bush temperatures increase greatly with increase in operating speed. According to graph (b) of Figure 9, the predicted and measured power losses lie closely together over the full range of speeds and the power loss increases significantly with speed. As graph (c) illustrates, compared with the experimental results the model moderately underestimates the flow rate, especially for lower speeds due to the neglect of the effect of film reformation. Both the predicted and measured flow rates increase considerably with speed. Graph (d) illustrates the predicted results of minimum film thickness. Clearly the predicted minimum film thickness increases moderately with speed.

4.3   Comparison with ESDU 84031

A complete design document for circular bearings, commonly known as the ESDU Item 84031 has already been developed. It is only valid for angular extents of grooves less than 40 degrees. Thus it is not strictly suitable to compare the current experimental results with the predictions of this document since the angular groove extent of the test bearing was over 50 degrees. Nevertheless, the current numerical model can be reliably checked with this design document.

Figures 10 and 11 show respectively the variations of the bearing performance with specific load and operating speed for an angular groove extent (ϕ) of 40 degrees. It can be seen that a fairly good agreement has been achieved between the predictions of the current model and the ESDU design scheme for all the performance characteristics presented over a wide range of operating conditions.

In fact, to some extent the groove size has only a modest effect on the bearing static performance according to the authors' experience. Hence it can be said that the current experimental results agree with the predictions of [1].

5   CONCLUSIONS

A theoretical and experimental investigation into the thermal behaviour of a two-axial-groove cylindrical bore bearing subjected to steady loading conditions has been presented and discussed. In the thermal analysis the effect of reverse flow has been taken into account and a robust oil groove mixing model has been developed to predict the inlet film temperature in a lobe. From this study, the following conclusions can be drawn:

- In comparing both the temperature profiles and the global performance characteristics, the theory and experiment have exhibited an excellent agreement over a wide range of loads and rotational speeds. This indicates that the current model is applicable to the thermal performance analysis and design of circular journal bearings.

- The maximum temperature, power loss and flow rate in the bearing are affected very significantly by rotational speed but only moderately by load.

- The effect of oil feed pressure has been examined and it has been shown that, except for the flow rate, the other static characteristics are hardly influenced by feed pressure.

- The predictions of the current model are in close correlation with those of [1] and the present experimental evidence supports the scheme of [1].

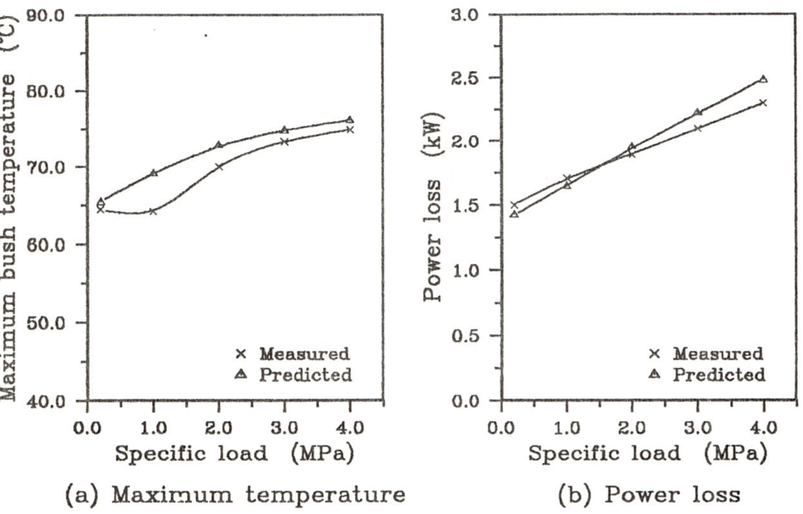

Figure 8 Variation of the bearing performance characteristics with specific load

N=4000 (rpm), ψ=0.0017, p_t=67 (kPa), T_t=45 (°C)

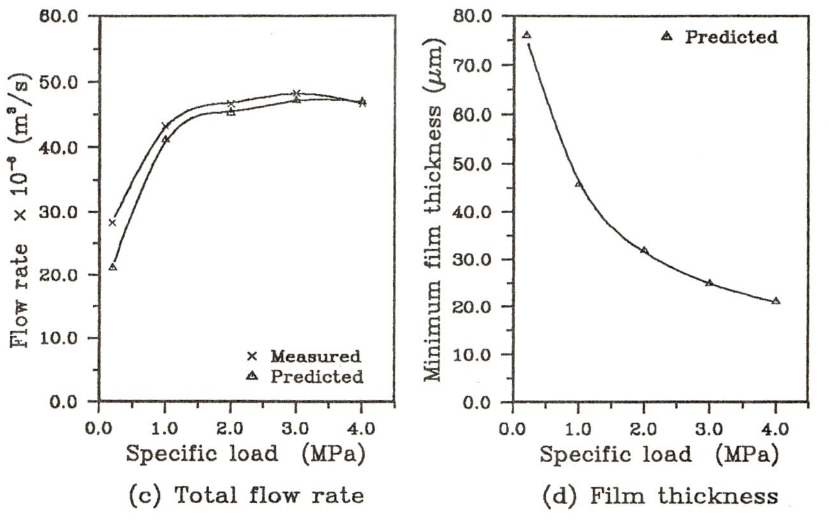

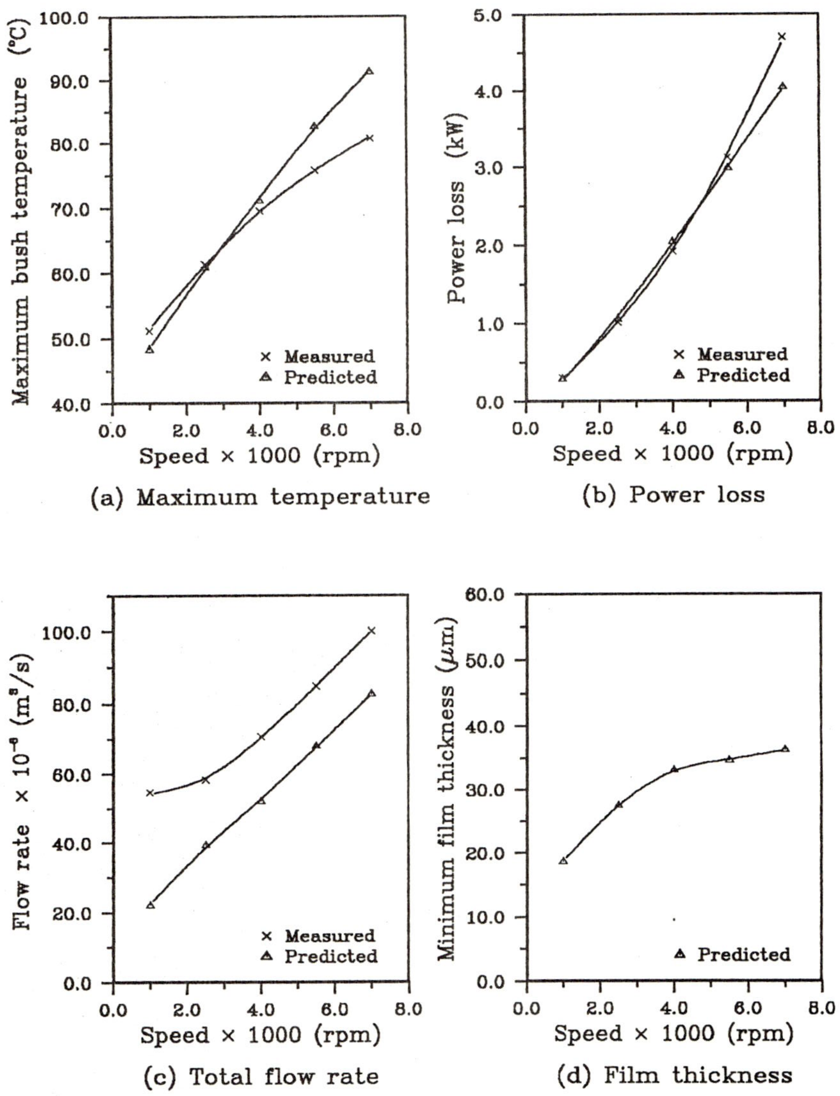

Figure 9 Variation of the bearing performance characteristics with rotational speed

W=16.94 (kN), ψ=0.0017, p_t=134 (kPa), T_t=45 (°C)

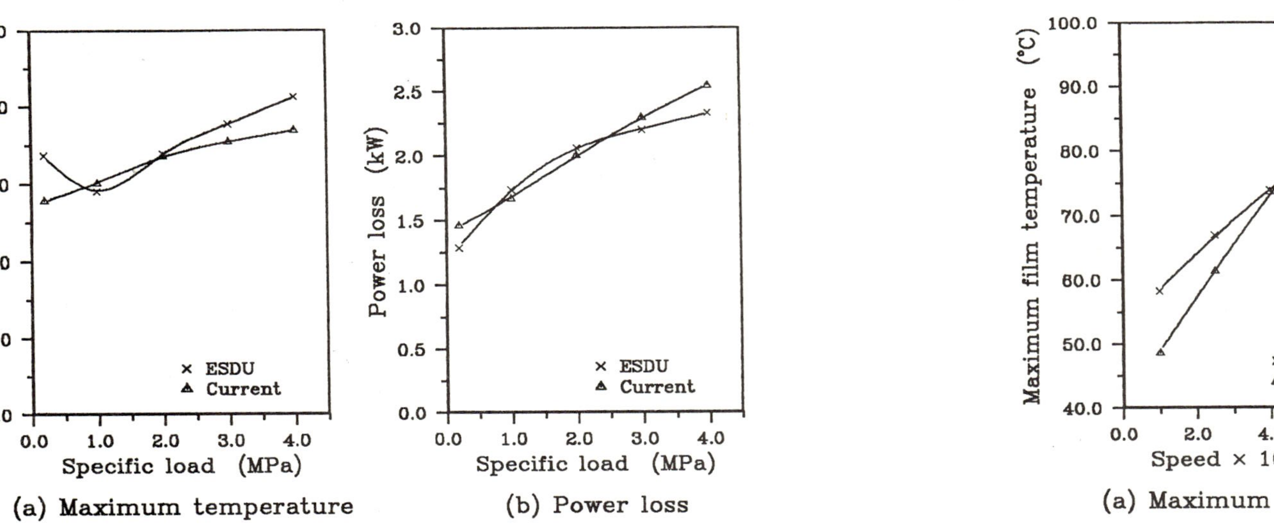

Figure 10 Comparison of predicted bearing performance
with current model and ESDU for different specific loads

N=4000 (rpm), ψ=0.0017, p_f=134 (kPa), T_f=45 (°C)

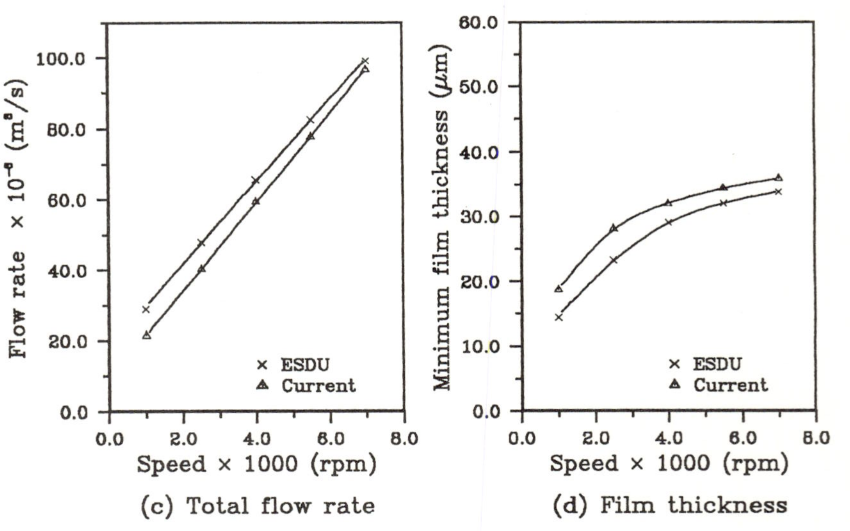

Figure 11 Comparison of predicted bearing performance
with current model and ESDU for different shaft speeds

W=16.94 (kN), ψ=0.0017, p_f=134 (kPa), T_f=45 (°C)

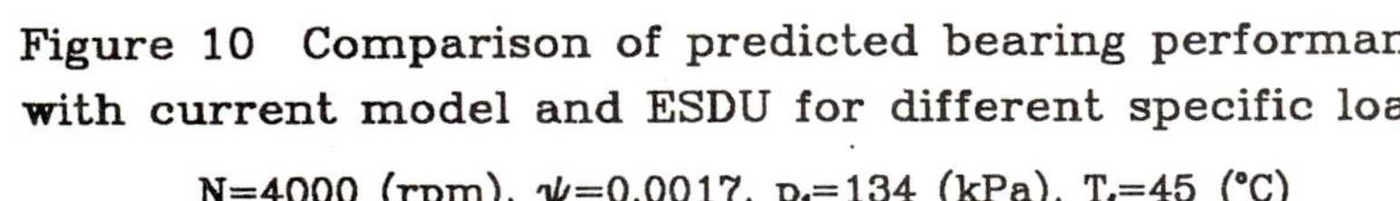

- Thermal distortion of the bush should be taken into account in the thermal analysis of a journal bearing at high rotational speeds.

- The computer program code developed based upon the current formulation is applicable to not only circular bearings but also various types of non-circular plain journal bearings. The detailed scheme and results for some types of non-circular bearing will be reported in future.

ACKNOWLEDGEMENTS

One of the authors (MTM) was supported by the Sir Y. K. Pao Foundation (Hong Kong), Chinese Education Commission and British Council under the Sino-British Friendship Scholarship Scheme. The research work described in this paper was part of a co-operative project with Michell Bearings - Vickers plc and the support of Mr S. Advani and Dr D. Horner is acknowledged with thanks. The test bearing was provided by the company. During the tests much help was given by Mr R. T. Harding, Mr L. Bellon, Mr A. Heald and other colleagues in the Department of Mechanical Engineering of Leeds University. All these are gratefully acknowledged.

References

[1] "Calculation methods for steadily loaded axial groove hydrodynamic journal bearings", *ESDU International plc, Data Item 84031, 1984.*

[2] **Khonsari, M. M.,** "A review of thermal effects in hydrodynamic bearings. Part II : Journal bearings", *Trans. ASLE,* Vol.30, 1987, p.26.

[3] **McCallion, H., Yousif, F. and Lloyd, T.,** "The analysis of thermal effects in a full journal bearing", *Trans. ASME J. Lub. Tech.,* Vol.92, 1970, p.578-587.

[4] **Singh, D. V., Sinhasan, R. and Prabhakaran Nair, K.,** "Elastothermohydrodynamic effects in elliptical bearings", *Trib. Int.,* Vol.22, 1989, p.43-49.

[5] **Knight, J. D. and Barrett, L. E.,** "Analysis of axially grooved journal bearings with heat transfer effects", *Trans. ASLE,* Vol.30, 1987, p.316-323.

[6] **Ott, H. H. and Paradissiadis, G.,** "Thermohydrodynamic analysis of journal bearings considering cavitation and reverse flow", *Trans. ASME J. Trib.,* Vol.110, 1988, p.439-447.

[7] **Lund, J. W. and Hansen, P. K.,** "An approximate analysis of the temperature conditions in a journal bearing. Part 1: Theory", *Trans. ASME J. Trib.,* Vol.106, 1984, p.228-236.

[8] **Gethin, D. T.,** "An investigation into plain journal bearing behaviour including thermo-elastic deformation of the bush", *Proc. I Mech E,* Vol.199, 1985, p.215-223.

[9] **Gethin, D.T. and El-Deihi, M.,** "Thermal behaviour of a twin axial groove bearing under varying loading direction", *Proc. I. Mech. E., Part C.,* Vol.204, 1990, p.77.

[10] **Ferron, J., Frene J. and Boncompain, R.,** "A study of the thermohydrodynamic performance of a plain journal bearing: Comparison between theory and experiments", *Trans. ASME J. Lub. Tech.,* Vol.105, 1983, p.422-428.

[11] **Boncompain, R., Fillon, M. and Frene, J.,** "Analysis of thermal effects in hydrodynamic bearings", *Trans. ASME J. Trib.,* Vol.108, 1986, p.219-224.

[12] **Mitsui, J., Hori, Y. and Tanaka, M.,** "Thermohydrodynamic analysis of cooling effect of supply oil in circular journal bearing", *Trans. ASME J. Lub. Tech.,* Vol.105, 1983, p.414-421.

[13] **Gethin, D. T.,** "A finite element approach to analysing thermohydrodynamic lubrication in journal bearings", *Trib. Int.,* Vol.21, 1988, p.67-75.

[14] **Suganami, T. and Szeri, A. Z.,** "A thermohydrodynamic analysis of journal bearings", *Trans. ASME J. Lub. Tech.,* Vol.101, 1979, p.21-27.

[15] **Sinhasan, R. and Chandrawat, H. N.,** "Analysis of a two-axial-groove journal bearing including thermoelastohydrodynamic effects", *Trib. Int.,* Vol.22, 1989, p.347-353.

[16] **Khonsari, K. K. and Wang S. H.,** "On the maximum temperature in double-layed journal bearings", *Trans. ASME J. Trib.,* Vol.113, 1991, p.464.

[17] **Tonneson, J. and Hansen, P. K.,** "Some experiments on the steady state characteristics of a cylindrical fluid-film bearing considering thermal effects", *Trans. ASME J. Lub. Tech.,* Vol.103, 1981, p.107.

[18] **Andrisano, A. O.,** "An experimental investigation on the rotating journal surface temperature distribution in a full circular bearing", *Trans. ASME J. Trib.,* Vol.110, 1988, p.638-645.

[19] **Gethin, D. T. and Medwell, J. O.,** "An experimental investigation into the thermohydrodynamic behavior of a high speed cylindrical bore journal bearing", *Trans. ASME J. Trib.,* Vol.107, 1985, p.538-543.

[20] **Mitsui, J., Hori, Y. and Tanaka, M.,** "An experimental investigation on the temperature distribution in circular journal bearings", *Trans. ASME J. Trib.,* Vol.108, 1986, p.621-627.

[21] **Ettles, C.** and **Cameron, A.**, "Considerations of flow across a bearing groove", *Trans. ASME, J. Lub. Tech.*, Vol.90, 1968, p.312.

[22] **Ettles, C.**, "Hot oil carry-over in thrust bearings", Proc. I. Mech. E., Vol.184, (Part 3L), 1970, p.75.

[23] **Heshmat, H.** and **Pinkus, O.**, "Mixing inlet temperature in hydrodynamic bearings", *Trans. ASME J. Trib.*, Vol.108, 1986, p.231-248.

[24] **Booser, E. R.** and **Wilcock, D. F.**, "Temperature fade in journal bearing exit regions", *Trans. STLE*, Vol.31, 1988, p.405-410.

[25] **Ma, M. T.**, THERMAL EFFECTS IN CIRCULAR AND NON-CIRCULAR PLAIN JOURNAL BEARINGS, *PhD thesis*, Department of Mechanical Engineering, University of Leeds (to be published), 1992.

APPENDIX 1 Estimation of Groove Pressure Flow

If the oil groove has no chamfers at the ends, the exit flow rate from the groove ends due to feed pressure can be approximately calculated by the following method. As Figure A1 shows, the ends of the groove can be considered as two dams. If the feed oil pressure is p_f and the width of the dams is a, the pressure gradient through a dam is linear and equal to $\frac{p_f}{a}$. The total exit

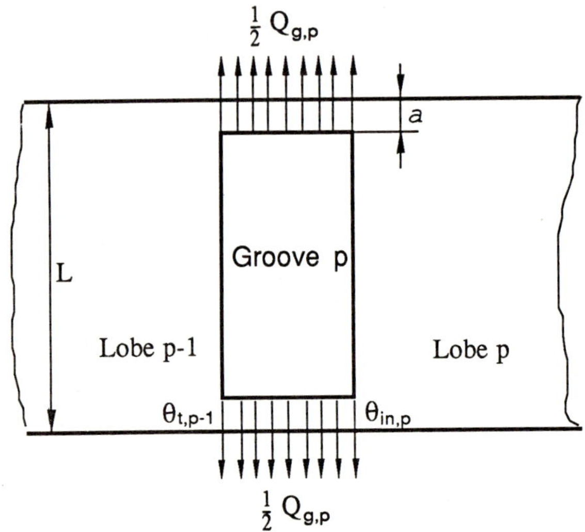

Figure A1 Exit flow from groove ends

flow rate from two ends of the groove can be calculated by

$$Q_{g,p} = 2 \int_{\theta_{t,p-1}}^{\theta_{in,p}} \frac{h^3}{12\eta_{g,p}} \frac{p_f}{a} R d\theta$$

where $\eta_{g,p}$ is the oil viscosity corresponding to the exit flow oil temperature $T_{g,p}$.

APPENDIX 2 Prediction of Temperature Fade

Instead of solving the energy equation, the temperature distribution in the cavitation region was predicted by an approximate method which included the cooling effect of the cold oil backflow from the downstream groove, as shown in Figure A2 (a). The scheme was based upon the following assumptions: (i) no energy dissipation; (ii) transverse temperature distribution is quadratic; (iii) divergent section is fully filled with lubricant. Thus the temperature distribution could be determined by applying boundary conditions (8), (9) and the elemental heat balance. The latter was established with the average transverse film temperatures and related the recirculating hot oil to the cold oil back flow (see Figure A2 (b)). The detailed derivation can be found in [25].

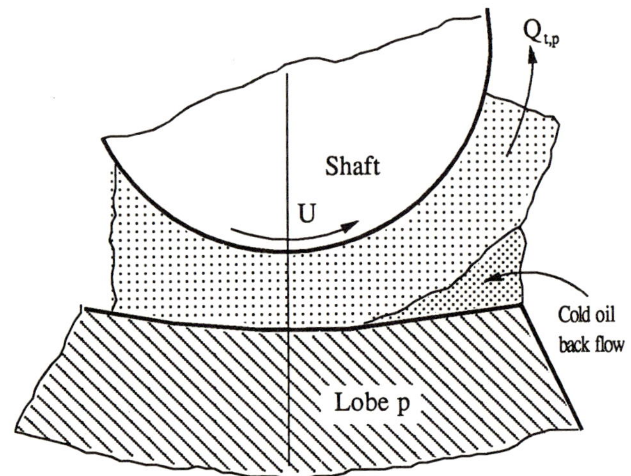

(a) Separation   model

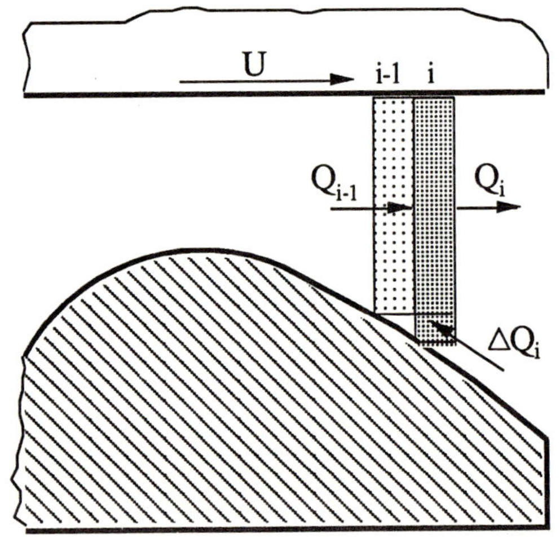

(b) Diagram of volume elements

Figure A2 Oil mixing in cavitation region

The influence of pivot location on the performance of tilting-pad thrust bearings

P B NEAL, BEng, PhD, FIMechE
Department of Mechanical and Process Engineering, University of Sheffield, UK
M A M SOLIMAN, BSc, PhD
Department of Production Engineering, Suez Canal University, Port Said, Egypt

SYNOPSIS The relative merits of central and offset pivot location in tilting pad thrust bearings has been a subject of practical interest since the time of Kingsbury and Michell. Test results are presented which show that the offsetting of the pivot offers benefits which, in certain cases, may be of crucial importance but that in many cases the penalties incurred in the use of centrally pivoted pads would be of little practical significance.

1 INTRODUCTION

According to classical iso-viscous hydrodynamic theory the generation of pressure in a thin fluid film requires that the film thickness decreases in the direction of sliding somewhere within the length of the film. The form of the reduction of film thickness is less important than its magnitude and, for a wide variety of film profile forms, a film ratio (ratio of inlet film thickness to minimum film thickness) of about 2 is optimal in terms of pressure generation. The necessary film convergence may range from the uniformly distributed convergence of the inclined plane pad to the localised stepped convergence of the Rayleigh pad.

If the convergence is achieved by machining the appropriate profile on the face of a fixed pad, the optimum film ratio is obtained for only one particular value of minimum film thickness and, therefore, for limited combinations of load and speed. The special merit of the inclined plane pad lies in the fact that the very small convergence required may be achieved by the beautifully simple alternative of pivoting the pad at its rear surface. In this way the film centre of pressure is automatically located and the film ratio is constrained to take up a particular value. Thus an appropriately located pivot ensures optimum performance for all operating conditions. Such reasoning led Michell to the invention of the tilting pad thrust bearing.

Michell's analysis (1) of the finite inclined plane pad showed that the centre of pressure would always be located in the trailing half of the pad. The optimum pivot location for a tilting pad is approximately 0.6 of the pad length from the inlet end of the pad and corresponds to a film ratio of about 2. A central pivot corresponds to a parallel film and a pad so pivoted should, therefore, have zero load capacity.

The idea of the tilting pad also occurred to Kingsbury as a result of reading Reynolds' paper. From autobiographical notes, written in 1942 and published posthumously in 1950 (2), it is evident that Kingsbury was aware that theoretically the centre of pressure would be downstream of pad centre. However, not liking 'a lopsided arrangement' and the limitation of only one direction of rotation that an offset pivot implied, Kingsbury built and successfully tested a bearing·with pads centrally pivoted on spherical bosses. Kingsbury's initial tests were conducted in 1898 but it was not until 1912 that his design was incorporated into a commercial machine and even then it was not without initial failure.

Whilst Kingsbury's bearing resulted from an intuitive appreciation of the import of Reynolds' analysis, Michell's design adhered more closely to the outcome of his own mathematical analysis in featuring a pivot downstream of pad centre. A bearing constructed to Michell's design was first used commercially in 1907 and, apparently, without any initial problems.

The segmental tilting pad bearing proved to be immensely superior, in terms of load capacity and power loss, to the plain annular form which it rapidly replaced. By 1914 Newbigin (3), in a paper aptly entitled "The problem of the thrust bearing", refers to 'considerable numbers of tilting pad bearings in service'. Newbigin observed that the pivot location can be varied within moderate limits but at the cost of increased friction. In the ensuing correspondence Kingsbury reported 'a curious result' that his experiments with centrally pivoted pads had produced load capacity 'fully as great' as with theoretically best pivot location but that the pad friction was much greater. Stoney's contribution to the discussion went even further to claim that experiments showed that pivot location was practically immaterial - 'it worked equally well whether it was centrally pivoted, as was proposed by Sir Charles Parsons, or put out of centre as Mr Michell did'.

That centrally pivoted pads work at all evidently requires attention to factors not considered by either Reynolds or Michell. Most significantly they had based their analyses on the assumption that the pad face would be rigid and truly plane and the lubricant viscosity would be constant throughout the film. In the discussion on Newbigin's paper, Michell referred to irregular pressure distributions obtained with pads pivoted at or upstream of pad centre and linked these experimental observations with pad non-rigidity and geometrical imperfections of the pad surface. The variable viscosity explanation for the operation of centrally pivoted pads was favoured by Gibson (4) in 1919 for the reason that the oil would be heated and 'becomes thinned in its passage across the pad, thus approximating to the desired wedge film action'.

Pad surface convexity along the direction of sliding has been shown by Raimondi & Boyd (5) to be very significant in producing an upstream shift of centre of pressure, thereby enabling a centrally pivoted pad to carry a load.

Similarly, a decreasing lubricant viscosity in the sliding direction produces the same effect but the influence is relatively weak. It has been shown by El Sahy (6) that a ratio of upstream to downstream viscosity of 2 (corresponding to a film temperature variation from 60°C to 84°C for an ISO VG32 oil) would be necessary to enable a centrally pivoted pad to operate with film thickness about 0.7 of that for a pad with optimally offset pivot, for the same load, and the pad friction would be about twice as large.

A systematic experimental study of the behaviour of pivoted pads was reported in 1954 by Kettleborough, Dudley & Baildon (7). The study was conducted on gunmetal pads (of area 0.59 x 10^{-3} m^2 and 35 deg subtended angle) pivoted on spherical buttons which could be adjusted to provide pivot location in the range 0.4 to 0.6 from the pad leading edge. Since the pads were button mounted they were free to tilt radially as well as circumferentially but the indentation, caused by the button pivot, raises some doubt as to the precise location of the effective pivot. Kettleborough et al found that in terms of film thickness the optimum pivot location was in the range 0.5 to 0.55 whilst 0.6 gave least friction. Successful operation was achieved with pivot location upstream of pad centre and the authors concluded that the explanation lay in the variation of viscosity along the film. In the estimation of film thickness it was necessary to assume that the pad remained plane as it tilted and the actual film shape could not be determined. In consequence, the Raimondi and Boyd surface curvature effect could not be assessed.

In 1957 results of tests on a much larger bearing (660 mm dia, 8 pads each of area 0.0186 m^2) were reported by de Guerin & Hall (8). Each pad was supported on a radially disposed step located at 0.6 from the pad leading edge but the test results led the authors to conclude that the pads did not actually tilt and that hydrodynamic wedge action resulted from thermal distortion and load deflection of the upstream unsupported part of the pads. Tests were also conducted with the step pivot centrally located but no results wre given beyond the very significant statements that in the range of specific load 15 - 45 bar (the common design range) 'performance was scarcely affected by this change in pivot position' and that at high loads (apparently 250 - 300 bar) 'centrally pivoted pads gave performance 15 per cent worse than for offset pads'. However, the criterion of performance was not stated.

Gardner (9) included some comparative results of centrally and offset button-pivoted pads (steel pads with babbitt face) in a paper primarily concerned with a comparison of different pad and pad-facing materials. For a 6-pad 150 mm dia bearing, the 0.6 offset pads ran at 8000 rev/min some 20 K cooler than centrally pivoted pads for specific loads in the range 7 - 35 bar. More recently, Gardner (10) has shown that an offset of 0.75 gives lowest pad maximum temperature.

The present paper concerns tests conducted on an 8-pad 150 mm dia bearing with the object of assessing the influence of pivot location on bearing performance in terms of operating temperature, film thickness and power loss. The pads were line pivoted. The same set of pads was used for all tests.

2 TEST RIG AND INSTRUMENTATION

A detailed description of the test rig is given elsewhere (11). A single test bearing is located within a test housing which is supported on roller bearings. The rotating thrust collar (the runner) is located on the drive shaft which is supported in plain journal bearings in a pair of support pedestals.

Oil is supplied to the test bearing at pad inner radius whence it flows radially over the runner face. Oil leakage from the housing and/or air induction into the housing are limited by floating seals (diametral clearance 0.05 mm) at the shaft. Oil is withdrawn from the top of the housing.

Load is applied to the test bearing via a hydrostatic loading plate by hydraulic pistons housed in the pedestal. The hydrostatic film permits measurement of the test bearing frictional torque by means of a torque arm fitted to the test bearing housing. The applied load is transmitted by the drive shaft to a reaction bearing enclosed within a second housing.

2.1 Test bearing

The test bearing configuration and pad details are shown in Fig 1. The steel pads were faced with whitemetal approximately 0.4 mm thick. The pads were surface ground as a set to remove any initial surface crowning to ensure truly plane pads of equal thickness.

The test pads were pivoted on radially disposed needle rollers (3.5 mm diameter) located within a cage. The pads were located by pad stops which were also used to clamp the roller cage. Circumferential adjustment of the cage ensured that the same pivot location was imposed on all the pads.

The pads were provided with thermocouples fitted in holes drilled parallel to the pad face. Thermocouples were also mounted in the runner 2 mm from its face and connected to a shaft-mounted slipring unit.

A 3 mm diameter capacitance probe was set into the runner face to scan the pads at mid-radius to indicate film thickness and film shape. The probe, press fitted into the runner, was finish-ground co-planar with the runner surface. The probe was connected into an a.c. bridge circuit via the slipring unit.

3 TEST RESULTS

The results presented relate to an oil feed rate of 9 l/min (2 imp gal/min) at a constant supply temperature 50 ^{o}C. Oil to ISO VG32 was used in the tests.

The speed range covered is 1500 - 5500 rev/min and the load range up to 29 kN total bearing load. The results quoted relate to a full complement 8-pad bearing for which the maximum load of 29 kN corresponds to a specific load of 3.1 MPa (450 psi). Additional tests were conducted with a reduced complement of 4 pads for which the maximum specific load was 6.2 MPa (900 psi).

3.1 Operating temperature

Pad maximum temperature is shown in Fig 2 to decrease as the pivot position is moved downstream of pad centre (Xp=0.5). At 5000 rev/min the maximum temperature was about 12 K lower for Xp=0.7 than for central pivot. The temperature advantage of offset pivot increases with speed but was found to be little influenced by load. Runner temperature , shown in Fig 3, also decreases with pivot offset but to a smaller extent than pad maximum temperature.

The higher pad and runner temperatures for centrally pivoted pads are not accompanied by a corresponding increase of oil temperature within the housing which was only marginally greater and reflected only the small increase in bearing friction.

For any given load and speed, the temperature rise along a pad was found to be little dependent upon pivot location (Fig 4). However, the actual temperature distribution along a pad was significantly affected by pivot location. Fig 5 illustrates typical temperature distributions; pad maximum temperature occurs well within the length of a centrally pivoted pad but at or near the end of the pad when the pivot is offset downstream of centre.

3.2 Film shape and minimum film thickness

Typical pad surface profiles (circumferential at mid-radius) are shown in Fig 5 and show increasing film ratio (inlet/minimum film thickness ratio) for offset pads. In all cases minimum film thickness is at the trailing end of the pad. The influence of pivot location on minimum film thickness is shown in Fig 6 and clearly shows the advantage of offsetting the pivot.

3.3 Power loss

Bearing friction was little influenced by pivot location. Power loss decreased with pivot offset to the extent shown in Fig 7. The advantage for offset pivots is nearly constant over the speed range but is less significant at high speed - probably because of the greater influence of churning losses at the higher speeds.

4 DISCUSSION

The pad profiles shown in Fig 5 clearly show the convexity to be expected as the result of temperature gradient through the pad thickness and bending about the line pivot due to film pressure.

The very much greater inlet film thickness for offset pads allows greater oil flow into the film and favours a lower film inlet temperature and lower pad operating temperature (as shown in Fig 5).

In the discussion of a previous paper (12), dealing with centrally pivoted pads, Neal associated the location of pad maximum temperature with the location of minimum film thickness although information of actual film shape was not then available. In the present tests the film shape plots show convergence over the full length of the pad in all cases and the fall in temperature along the trailing part of centrally pivoted pads is clearly not associated with film divergence. There is evidence (13) that heat transfer from the film to the runner is of increasing importance as film thickness decreases. It may be seen in Fig 5 that the centrally pivoted pad operates with a much more extensive thin film region than do pads with offset pivots and heat transfer to the runner would therefore be expected to show its effect over a more extensive part of the film length. The fall in temperature towards the downstream end of the film requires only that the rate of heat transfer to the runner exceed the rate of energy dissipation within the film. A downstream fall in film temperature would be expected to result in a corresponding fall in pad temperature. A local drop in temperature at the trailing end of the pad (beyond the location of the fitted thermocouples) is to be expected due to exposure of the pad to the cooler oil in the housing but this applies to all pivot locations and seems unlikely to provide an alternative explanation for the behaviour of the centrally pivoted pads.

Within the range of pivot location tested, pad maximum temperaure decreased with pivot offset. However, film ratio increases with the amount of offset with the undesirable consequence of increased ratio of peak film pressure to mean pressure. Whilst maximum temperature at the pad surface is arguably the most important criterion of acceptable performance, since an excessive temperature will endanger the integrity of the whitemetal facing, it is the local combination of high temperature and high film pressure that is important. In this respect the less peaky pressure distribution with reduced pivot offset may have the advantage notwithstanding higher maximum temperature. Apart from this, there is no merit in seeking to achieve a lower temperature in circumstances in which pad temperature is entirely acceptable.

In situations in which operating film thickness is the critical factor, there is clear advantage with offset pivot. However there is no merit in designing for film thickness larger than is necessary to avoid metal-to-metal contact and to provide for overload capacity. In terms of minimum film thickness the optimum pivot location is clearly with offset of around 0.6 to 0.65 and, in view of the comment above concerning the local combination of pad temperature and film pressure, probably approximates to the overall optimum. This is very much in line with the results of Michell's analysis notwithstanding the constant viscosity inclined plane basis of that analysis.

For centrally pivoted pads , depending as they do on deflected pad form, mathematical analysis of the general case is not possible. It says much for Kingsbury's intuition that centrally pivoted pads provide a perfectly adequate answer to many thrust bearing requirements.

5 CONCLUSION

The test results presented show that pivot offset downstream of pad centre provides benefits which in some cases may be of crucial importance but that in many cases the penalties incurred in the use of centrally pivoted pads would be of little practical significance. In situations involving bi-directional operation the centrally pivoted pad represents the obvious and entirely adequate practical compromise.

REFERENCES

(1) Michell, A.G.M. The lubrication of plane surfaces. Zeit Math und Physik. 1905, 52, 123-137.

(2) Kingsbury, A. Development of the Kingsbury thrust bearing. Mechanical Engineering. 1950, 72, 957-963.

(3) Newbigin, H.T. The problem of the thrust bearing. Proc I C E 1914, 196, 223-246.

(4) Gibson, J.H. The Michell thrust block. Engineering, 1919, 765-768.

(5)   Raimondi, A.A. and Boyd, J. The influence of surface profile on the load capacity of thrust bearings with centrally pivoted pads. Trans ASME. 1955, 77, 321-330

(6)   El Sahy, T.I.A. A computer based study of pad thrust bearings. M Eng thesis, University of Sheffield, 1986.

(7)   Kettleborough, C.F., Dudley, B.R. and Baildon, E. Michell bearing lubrication. Proc I Mech E, 1955, 169, 746-759.

(8)   de Guerin, D. & Hall, L. F. Some characteristics of conventional tilting-pad thrust bearings. I Mech E Conf on lubrication and wear. 1957, paper 82.

(9)   Gardner, W. W. Performance tests on six-inch tilting pad thrust bearings. Trans ASME Jnl Lub Technology, 1975, 97, 430-438.

(10) Gardner, W. W. Tilting pad thrust bearing tests - influence of pivot location. ASME Paper No. 88-Trib-3, 1988.

(11) Soliman, M. A. M. Performance characteristics of pivoted pad thrust bearings. Ph.D. thesis, University of Sheffield, 1986.

(12) Neal, P. B. Some factors influencing the operating temperature of pad thrust bearings. 6th Leeds-Lyon Symposium on Tribology, 1979, 137-143.

(13) Neal, P. B. Heat transfer in pad thrust bearings. Proc I Mech E, 1982, 196, 217-228.

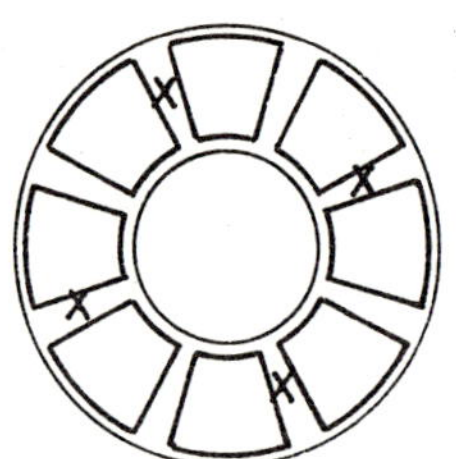
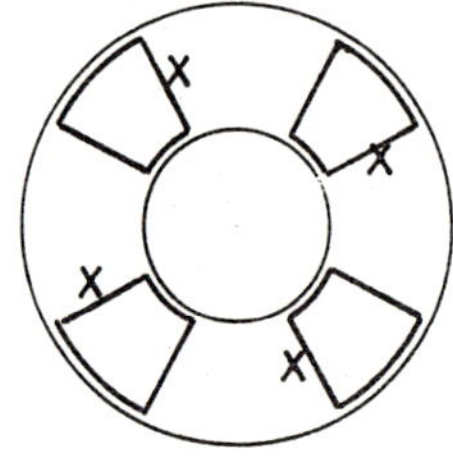

x   Thermocouple locations

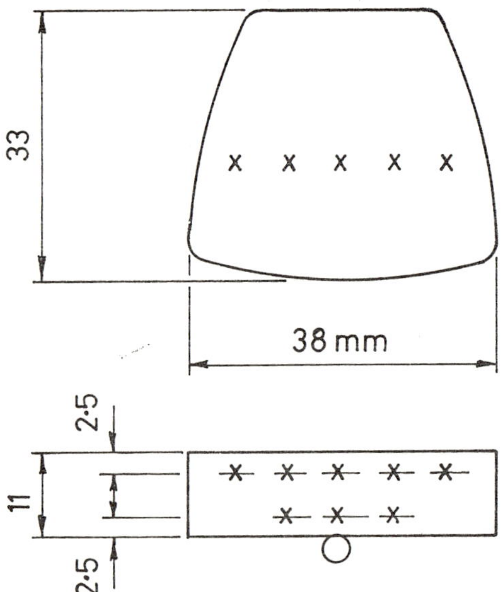

Bearing mean diameter 116 mm

Pad area 1150 mm^2

Fig 1   Test bearing configurations and pad details

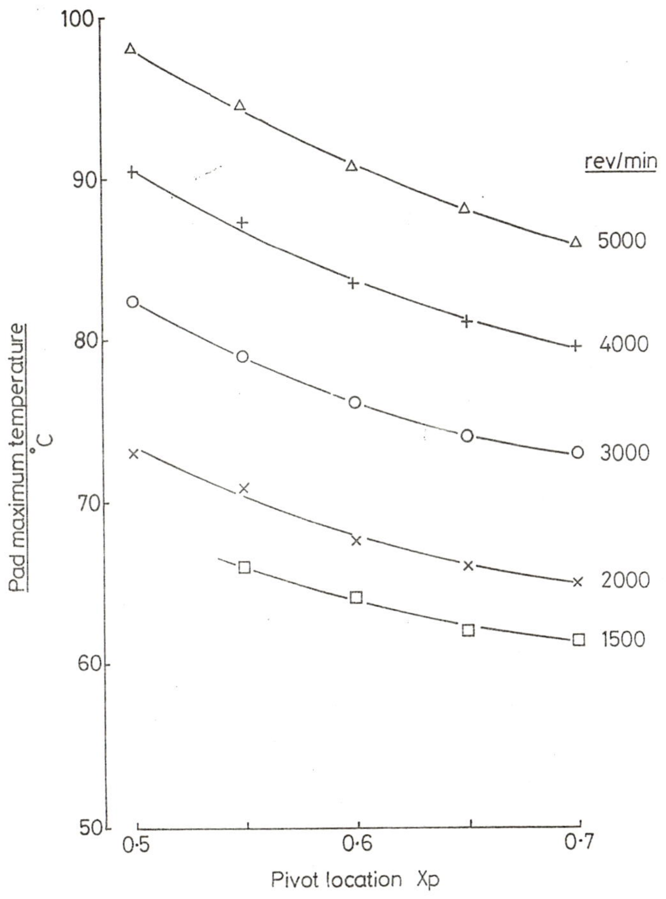

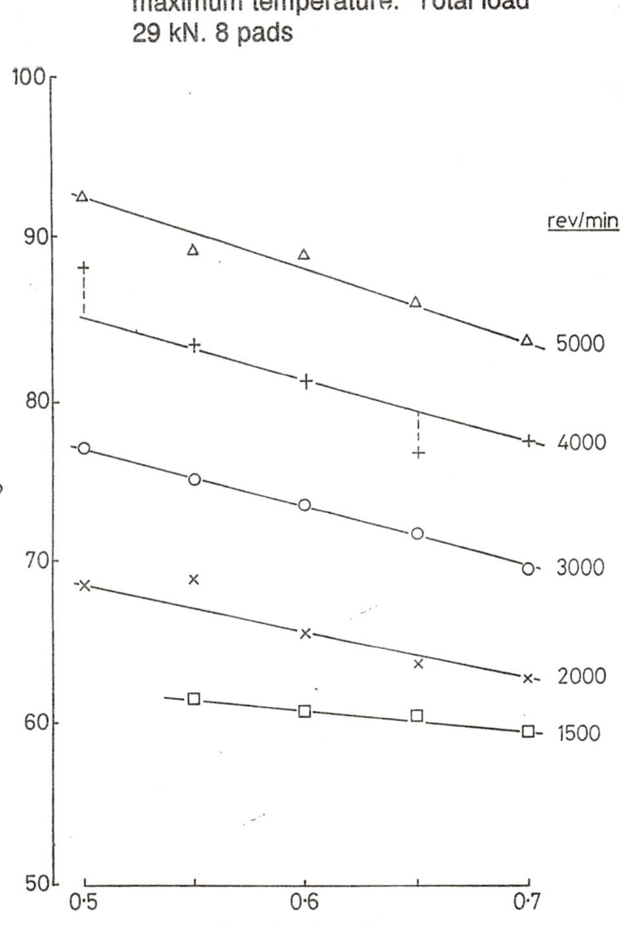

Fig 2 Influence of pivot location on pad maximum temperature. Total load 29 kN. 8 pads

Fig 3 Influence of pivot location on runner temperature. Total load 29 kN. 8 pads

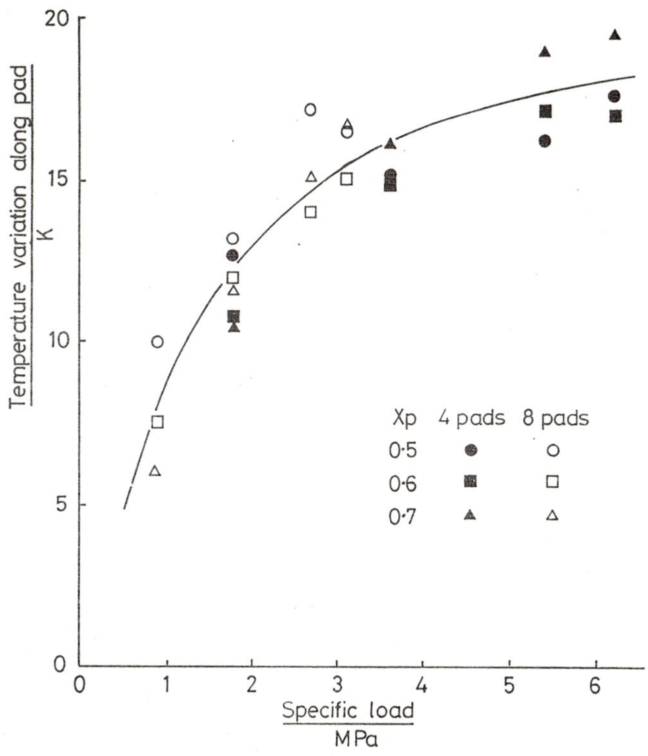

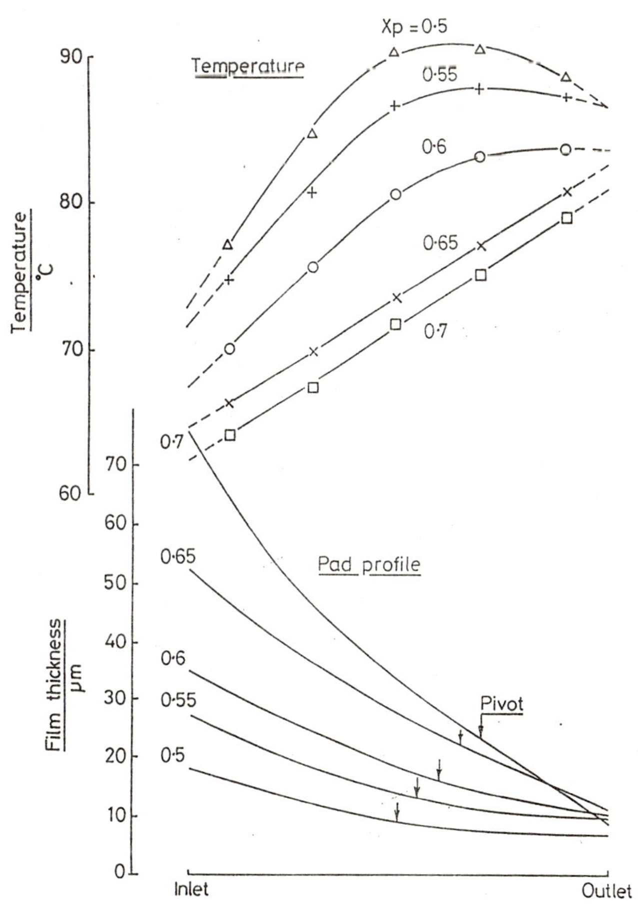

Fig 4 Temperature rise along pad. Speed 5000 rev/min

Fig 5 Pad surface temperature distribution and pad profile. (Same pad in each case.) Specific load 3.1 MPa. Speed 4000 rev/min

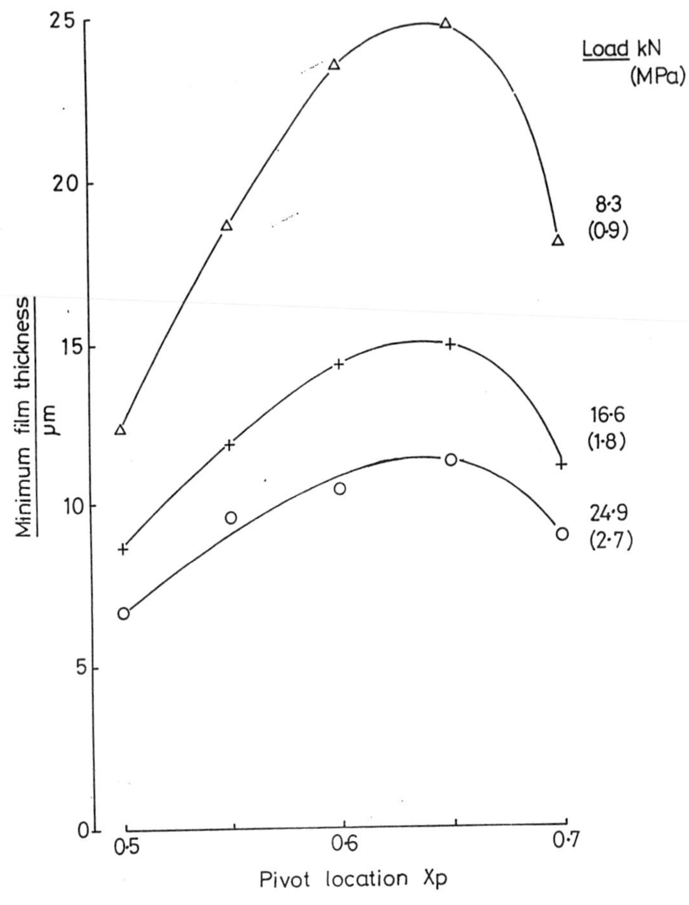

Fig 6   Influence of pivot location on minimum film thickness. Speed 4000 rev/min

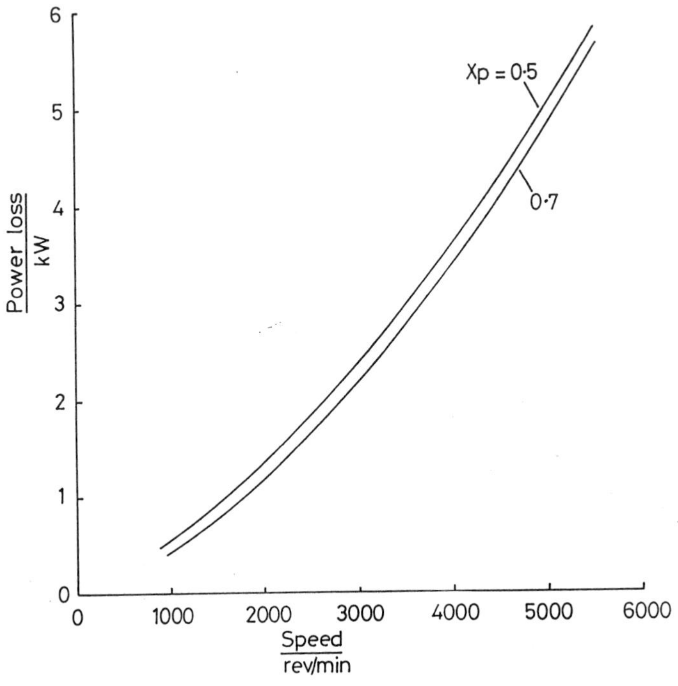

Fig 7   Bearing power loss. Load 29 kN. 8 pads

Performance evaluation of the LEG tilting pad journal bearing

K BROCKWELL, MSc, DIC, CEng, MemSTLE, MIMechE and W DMOCHOWSKI, MSc, PhD
National Research Council Canada, Vancouver, Canada
S DeCAMILLO, BSc, MemSTLE, MamASME and A MIKULA, MBA, BSc, MemSTLE, MemASME
Kinsbury Inc., USA

SYNOPSIS This paper discusses the performance characteristics of the leading-edge-groove (LEG) tilting pad journal bearing and presents new experimental temperature data from a bearing operating with "on pad" and "between pads" loading conditions, and with different bearing clearances. This data is then compared with results obtained from a computer model of the LEG bearing. Good correlation between the theoretical and experimental results suggests that the technique used in the model to calculate the temperature of the oil at the leading edge of the LEG journal pad is reasonably accurate.

The experimental data is collected from a 0.098 m diameter, five pad bearing, operating at shaft speeds up to 16500 rev/min and with unit loads up to 3 MN/m^2. The paper also presents experimental results from bidirectional testing of the offset pivot LEG bearing, and finally discusses a technique for achieving substantial reductions in the power loss of the LEG bearing.

1 INTRODUCTION

The circular bore journal bearing may experience self excited subsynchronous vibration during operation, and although other bore modifications such as the multi-lobe and offset halves designs are successful at raising the stability threshold, only the tilting pad journal bearing offers the possibility for eliminating oil film instability. These unique characteristics were confirmed by Brockwell *et al.* (1) in a series of experiments on a conventional tilting pad bearing. They found that the cross-coupled coefficients were negligible in comparison to the direct coefficients, providing that there was geometrical and "thermal" symmetry of the bearing. Ettles (2,3) considered the effect of lack of symmetry on the performance of the tilting pad journal bearing.

Because the tilting pad journal bearing is finding increased usage in high power density machinery, interest is focusing on the bearing's steady state performance. For example, Brockwell and Dmochowski (4) found that, during operation, significant changes to the bearing and pad clearances may lead to a pad preload that is substantially different to that specified in the initial design.

In recent years, a new design of hydrodynamic bearing, known as the leading-edge-groove (LEG) tilting pad bearing, has been the subject of further development work. The LEG bearing is so named because the leading edge of each pad is extended to accommodate an axial oil distribution groove that directs a controlled amount of cool lubricant into the hydrodynamic oil film. An earlier experimental study (5) focused on a thrust version of the LEG bearing, when tests on a 267 mm outside diameter bearing at speeds up to 13000 rev/min indicated significant reductions in frictional loss and bearing operating temperature. More recently, a journal version

of the LEG tilting pad bearing was the subject of preliminary experimental and theoretical studies (6). This work showed that the LEG bearing had significantly lower operating temperatures to those of the conventional bearing; a characteristic that was attributed to the reduction in hot oil carry over between one pad and the next.

This paper is an extension of the work described by Dmochowski *et al.* (6) and presents new experimental temperature data from the LEG journal bearing for both "on pad" and "between pads" loading conditions, and for different bearing clearances. The experimental results are compared with data from a computer model of the LEG bearing. Furthermore, the paper discusses results obtained from bidirectional testing of the LEG bearing, and describes a technique for reducing the power loss of the LEG bearing.

2 EXPERIMENTAL APPARATUS

2.1 <u>Description of test rig</u>

The test facility is described in detail elsewhere (7). Briefly, the 0.098 m diameter shaft is supported on high precision, angular contact ball bearings, and driven by a variable speed electric motor through a belt-pulley system to give a range of speeds between 1800 and 16500 rev/min. The test bearing is mounted in a special housing located midway between the support bearings and moves in response to a vertical load that is applied to the housing by a tensioned cable. Eddy current probes mounted on the ends of the housing measure the horizontal and vertical displacements of the bearing and provide a check of the alignment of the bearing with respect to the shaft. Three axial tensioned wires attached to each end of the housing minimize non-parallel movement of the bearing with

Table 1   Bearing and pad clearances

Bearing group	Bearing radial clearance	Pad radial clearance	Preload
	mm	mm	
1	0.076	0.102	0.25
2	0.102	0.102	0.00
3	0.051	0.102	0.50

respect to the shaft. Because these wires have lateral flexibility, their influence on the radial movement of the bearing housing is small.

The lubricant used in this study was a light turbine oil (ISO VG32) with a viscosity of 0.02325 Pa.s at 40C and 0.0054 Pa.s at 100C. The lubrication system incorporated a feedback control device and a shell and tube heat exchanger to regulate the oil supply temperature to the bearing. The inlet temperature was maintained between 48C and 50C.

Bearing circumferential temperature distributions were measured by copper-constantan thermocouples imbedded in the babbitt lining to within 0.5 mm of the bearing surface (Figure 1 shows the location of these thermocouples). These measurements, together with all other data from the rig, were stored in the memory of a data acquisition system and passed to a host computer at a later time.

2.2 <u>Description of test bearings</u>

The bearings have a nominal diameter of 0.098 m, a length /diameter ratio of 0.387 and consist of five babbitt lined journal pads supported in a precisely machined aligning ring. This ring is split axially to allow easy assembly of the bearing around the shaft. An annular oil distribution groove is machined into the outside of the aligning ring and, in the case of the conventional bearing, oil is directed from this annulus through radial feed holes into the spaces between adjacent pads. Cool oil to the pads of the LEG bearing is directed to the centre of the leading edge grooves by small diameter pipes fitted with O rings (see Figure 2). The pads of both bearings have an effective angle of 56.1 degrees, although the overall angle of the LEG pad is somewhat larger to accommodate the leading edge groove. The axial length of this groove is 33.5 mm. To allow the pads to adjust to conditions of axial misalignment, their back surface is contoured both circumferentially and axially. The shoes are held axially and circumferentially by retaining plates. Details of bearing and pad clearances, and resulting pad preload values, are given in Table 1.

Each group comprised a 0.6 offset pivot LEG bearing. In addition, group 1 included a 0.6 offset pivot conventional bearing.

2.3 <u>Bearing test conditions</u>

Tests were conducted with loads of 5.18 kN and 11.1 kN, shaft speeds that ranged between 1800 rev/min and 16500 rev/min and nominal flow rates that depended on the actual test condition (see Table 2). The group 2 LEG bearing was tested for both "on pad" (LOP) and "between pads" (LBP) loading configurations. Tests were also performed on bearing groups 1, 2 and 3 to determine the effect, on bearing temperature, of pad radial clearance. Finally, the group 1 LEG bearing was run in reverse rotation.

3 BASIS OF THE COMPUTER MODELS

Computer models that calculate the performance of the conventional and LEG tilting pad journal bearings have been developed and are described elsewhere (6). Briefly, the pressure distribution in the oil film is calculated from a two-dimensional version of the Reynolds' equation that considers viscosity variations in the circumferential and radial directions, and assumes the Swift-Steiber boundary condition for a cavitated oil film. Turbulence is accounted for in a manner similar to that described by Constantinescu (8) and Frene and Constantinescu (9). The temperature distribution in the oil film is governed by the energy equation, accounting for heat conduction across the oil film and heat convection in the circumferential direction. The energy equation is solved assuming the following boundary conditions: 1) a shaft temperature that is equal to the average of the pad temperature distributions; 2) a certain temperature distribution at the pad surface (calculated from the heat transfer equation); and 3) a pad leading edge temperature that is constant through the thickness of the oil film. The third boundary condition is calculated on the basis that hot oil from the preceding pad affects the oil temperature at the leading edge of the next pad. This temperature is calculated from simple heat balance equations which represent pad leading edge conditions in both the conventional and LEG bearings. This matter is discussed later in the paper.

The temperature distribution at the pad surface is calculated from the Laplace heat transfer equation, assuming a constant temperature in the axial direction. The boundary conditions for the heat transfer equation correspond to an equality of heat fluxes at the boundary between the oil film and the pad surface, and heat

Table 2   Nominal oil flow rates, 10^{-4} m^3/s

Load	Shaft speed					
N	rpm					
	1800	3600	5000	9000	12000	16500
1300	0.15	0.39	0.61	1.37	2.11	3.60
5175	0.21	0.52	0.76	1.56	2.32	3.81
10350	0.28	0.66	0.93	1.82	2.63	4.21
11100	0.21	0.65	0.95	1.85	2.67	4.27

convection on all remaining surfaces. Approximate pad deflection is calculated from the one-dimensional equation for a beam. As in (2), shear forces, bending moments and local differences in temperature across the thickness of the pad are taken into consideration. In the axial direction, the load distribution is assumed constant and is calculated by averaging the pressures in the axial direction. Boundary conditions at the pivot correspond to zero pad surface deflection and zero gradient.

By simultaneously solving these equations for each individual pad, the solution for the entire bearing is obtained in a manner similar to that originally proposed by Ettles (2).

4 PERFORMANCE EVALUATION OF THE LEG BEARING

4.1 Bearing operating temperatures

Dmochowski *et. al.* (6) discuss some preliminary tests of the group 1 LEG bearing and compare it's performance with that of the group 1 conventional tilting pad journal bearing. Both bearings had a pivot location of 0.6. These tests showed that, at shaft speeds above 12000 rev/min, the LEG bearing had substantially lower operating temperatures. For example, at 16500 rev/min with a load of 11.1 kN, the measured maximum temperature of the LEG bearing was 105C, compared to 127C in the case of the conventional bearing. When the performance characteristics of the bearings were analyzed using the computer models described in this paper, it was concluded that the lower operating temperatures of the LEG bearing were the result of reducing the effect of hot oil carry over. Furthermore, it was found that simple heat balance equations could be used when calculating pad leading edge oil temperatures. In the case of the conventional bearing, the authors assumed that all of the flow q_2 (at temperature t_2) leaving the pad trailing edge will enter the leading edge of the next adjacent pad (the limitation is that q_2 cannot exceed 90 per cent of q_i). Thus:

$$t_1 = \frac{(q_1 - q_2)t_i + q_2 t_2}{q_1} \qquad (1)$$

where:

t_1 = oil temperature at the leading edge
t_2 = oil temperature at the trailing edge of the preceding pad
t_i = oil inlet temperature to the bearing
q_1 = oil flow at the leading edge
q_2 = oil flow at the trailing edge of the preceding pad
q_i = oil flow to each pad (total flow/number of pads)

In the case of the LEG bearing, it was assumed that all cold oil supplied to the leading edge groove enters the film at the leading edge of the pad, with the balance of the flow being provided by the hot oil carried over from the previous pad. Hence:

$$t_1 = \frac{q_i t_i + (q_1 - q_i)t_2}{q_1} \qquad (2)$$

Equations (1) and (2), when used in conjunction with the computer models described in this paper, were found to give calculated bearing temperatures that were in good agreement with the measured temperatures (6).

The work of Dmochowski *et. al.* (6) also featured an investigation of the effect of oil flow rate on bearing temperatures. In the case of the conventional bearing, it was found that halving or doubling the flow rate had little effect on the operating temperature of the bearing, probably because there was always enough oil in the cavity of the bearing to provide the right amount of "make up" flow (q_1) at the leading edge of each pad, even when the oil flow rate was reduced by 50 per cent. Consequently, variations in pad operating temperature were small.   On the other hand, the LEG bearing was more sensitive to oil flow, with the result that this bearing had a different operating characteristic from that of the conventional bearing. These tests showed that the operating temperature of the LEG bearing rose slightly when the oil flow was reduced by 50 per cent. Under such operating conditions, it would seem that less cool oil from the oil supply groove, and more hot oil from the previous pad, now enters the oil film, resulting in higher pad temperatures.

4.2 LOP vs. LBP bearing tests

This section of the investigation featured "LOP" and "LBP" tests on the group 2 LEG bearing with loads of 5.18 kN and 11.1 kN, and with shaft speeds that ranged between 1800 and 16500 rev/min. Figure 3 shows measured bearing temperatures from these tests. With regard to the results for the 5.18 kN load (Figure 3(a)), it is of interest to note that bearing temperatures are similar for both loading configurations with shaft speeds up to 16500 rev/min. This is also true of the results from the 11.1 kN load tests for shaft speeds up to approximately 12000 rev/min, although at higher speeds there are differences. For example, at 16500 rev/min, the measured difference was 8C, with the "LOP" operating at the higher maximum temperature of 116C. The results seem to suggest that , for less severe operating conditions, the bottom loaded pad(s) of the bearing operate at a similar temperature, regardless of whether the load is "on" or "between" pads. However, for more severe operating conditions when the load is "on pad", the increase in bearing temperature is thought to be the result of a thinner operating oil film.

Figure 3 also presents calculated results. These results are in fairly close agreement with the measured bearing temperatures and suggest that the assumptions made in the model (particularly the boundary conditions for the energy equation) are quite reasonable.

4.3 Effect of bearing clearance

Tests were conducted on bearing groups 1 to 3 inclusive to examine the effect of bearing clearance on the operating temperature of the LEG bearing. It should be noted that it was the "bearing" clearance that was changed for this series of tests. Figure 4 presents measured and calculated maximum bearing temperatures for the three bearing groups. The bearings were tested with both load "on pad" and "between pads", for a range of shaft speeds up to 16500 rev/min and a load of 11.1 kN.

Figure 4 has a number of interesting features. Firstly, as with the results presented in the previous section, differences in temperature between LOP and LBP were only noticeable at the higher shaft speeds. This was true of all three bearings. Secondly, reducing the "bearing" clearance from 0.102 mm to 0.076 mm had little effect on bearing operating temperatures, although when the clearance was further reduced to 0.051 mm, quite significant increases in temperature were observed. Since this increase in bearing temperature was particularly rapid at higher shaft speeds, tests on the group 3 bearing were restricted to a maximum speed of 12000 rev/min. Unfortunately, LOP tests on the group 3 bearing were not conducted.

4.4 Bidirectional tests

The leading-edge-groove tilting pad journal bearing features an offset pivot and a leading edge oil distribution groove that are intended to improve the performance of the bearing, but, it is believed, make it suitable for only one direction of rotation. Certain machines, however, require a bearing that is suitable for bi-directional operation. These machines usually operate in one direction for much of their operational life, with only short excursions in the reverse direction of rotation. The conventional centre pivoted pad is suitable for both directions of rotation and is often used in such applications. However, when compared to the performance characteristics of the offset pivoted bearing, the centre pivoted bearing is usually found to operate with a higher babbitt temperature and a thinner oil film.

The purpose of this experimental study was to show that the LEG tilting pad journal bearing is suited to bidirectional operation, and that it can operate in reverse rotation with the maximum rated load of the bearing without causing damage to the babbitt surfaces. Bidirectional testing of the LEG bearing therefore resulted in the pivot being located in the leading half of the pad at the 40 per cent position, and the oil supply groove being positioned at the trailing edge of the pad. Because the position of the thermocouples was unchanged from the "forward" rotation tests, it was thought that this arrangement would not be capable of detecting the maximum temperature of the bearing when operating in reverse. However, it should be possible to estimate the maximum temperature of the bearing and Figure 5 shows how this was done. Essentially, this involved estimating the pad temperature gradient in the direction of rotation from the three measured temperatures on the leading half of the pad and assuming that the maximum temperature will occur at about the 60 per cent location. This would seem to be a reasonable assumption given that the oil film does not extend to the end of the pad, but ends, probably, somewhere in the vicinity of the 60 per cent location.

Figure 6 shows maximum pad temperatures for both the forward and reverse directions of rotation, with a load of 11.1 kN and shaft speeds that ranged between 1800 and 16500 rev/min. As expected, bearing temperatures were higher when running in reverse rotation, although the difference was not that significant at speeds up to 6000 rev/min. At higher speeds the difference was more noticeable, and was approximately 10C at 12000 rev/min. Results from tests with a load of 5.18 kN, while not shown on Figure 6, indicated a maximum temperature of only 95 C at 12000 rev/min, suggesting that higher speeds at lower loads should be possible.

At the conclusion of the reverse rotation tests, the bearing was found to be in a sound condition and suitable for further operation.

Table 3   Calculated power losses and maximum pad temperatures. 11.1 kN load "between pads". Group 1 bearings. Shaft speed = 16500 rpm.

| Bearing | Power loss | | | | | | Max. pad temp. |
| | kW | | | | | | °C |
	pad 1	pad 2	pad 3	pad 4	pad 5	total	
conventional	3.48	4.48	1.88	1.55	1.85	13.24	124
LEG	3.36	4.03	2.39	1.72	1.87	13.37	107
LEG (T_i=60°C)	2.90	3.33	1.61	1.16	1.39	10.39	111
LEG (T_i=74°C)	2.36	2.53	1.05	1.19	0.97	8.1	121

5 REDUCING THE POWER LOSS OF THE LEG BEARING

The authors were unable to measure accurately the friction torque of the test bearings, but instead used their computer models to analyze the power loss of the bearings. The purpose of this work was twofold: 1) to identify a method for reducing the power loss of the LEG bearing; and 2) to draw a comparison between the losses of the LEG and conventional bearings. Figure 7a shows that the calculated losses of the LEG and conventional bearings, for an oil inlet temperature of 49C, are similar. However, the maximum operating temperature of the LEG bearing is significantly lower (107C compared to 124C for the conventional bearing). Therefore, it should be possible to raise the oil inlet temperature of the LEG bearing so as to reduce it's power loss, and maintain a maximum bearing temperature that is equal to, or lower than, that of the conventional bearing. As an extreme example, Figure 7a shows calculated power losses for the group 1 LEG bearing where the inlet temperature is raised to the point where the maximum temperature of the LEG and conventional bearings are similar. (Obviously, the rise in inlet temperature is dependent on bearing operating conditions, and details on other shaft speeds are given in Figure 7b). At 16500 rev/min, with a load of 11.1 kN, the oil feed temperature to the LEG bearing must be raised to 74C to give a maximum temperature similar to that of the conventional bearing (with an inlet temperature of 49C). Figure 7a shows that substantial power savings are possible, particularly at higher shaft speeds; for example, with an inlet temperature of 74C the savings are of the order of 40 per cent at 16500 rev/min. Table 3 shows the losses of the individual pads of the conventional and LEG bearings for an inlet temperature of 49C, as well as those of the LEG bearing with elevated inlet temperatures of 60C and 74C. It is worth noting that the minimum film thickness is still acceptable with a 74C inlet temperature (i.e. $1.8*10^{-5}$ m compared to $2.4*10^{-5}$ m with a 49C inlet temperature).

Table 3 also lists power losses for a 60C inlet temperature, as 74C might be considered a rather extreme condition. For this more moderate rise in inlet temperature (11C), the savings in power loss are still substantial (approximately 22 per cent), and the maximum pad temperature is only 111C (still 13C lower than the conventional bearing with a 49C inlet temperature). Table 3 shows that, as a result of increasing the oil inlet temperature, significant savings in power loss are achieved at all pads.

Figure 8 shows the calculated pad temperature distributions of the conventional and LEG bearings. This figure has two interesting features. Firstly, raising the inlet temperature to the LEG bearing has most effect on the operating temperature of the unloaded pads (3, 4 and 5). Since little heat is, in fact, generated at the unloaded pads, this dramatic increase in temperature is probably the result of introducing hot oil at the leading edges of the pads, and not the consequence of hot oil carry over. Secondly, the leading edge temperatures, and therefore the operating temperatures, of the loaded pads of the conventional bearing are similar to those of the LEG bearing (with the raised inlet temperature). In the case of the conventional bearing, leading edge oil temperatures are strongly influenced by hot oil carry over from the preceding pads (see equation 1). However, the effect of hot oil carry over is only minimal in the case of the LEG bearing, although the oil supply temperature is higher. The overall result is that, when comparing LEG and conventional bearing temperatures, differences on the loaded pads are less significant than those of the unloaded pads.

The elevated operating temperature of the unloaded pads has two effects on bearing performance. Firstly, power loss in the upper half of the bearing is lowered as a result of reducing viscous shearing losses. Secondly, the increased temperature gives rise to a hotter running shaft and this translates into further reductions in power loss in the bottom half of the bearing as oil film temperature profiles are modified. In the example given in Figure 8, the calculated shaft temperature for the LEG bearing is 68C with an inlet temperature of 49C. This increases to 85C when the inlet temperature is raised to 74C.

6 CONCLUSIONS

Testing of a 0.098 m diameter five shoe, tilting pad journal bearing, fitted with a leading-edge-groove lubricant supply method, at loads of 5.18 kN and 11.1 kN and shaft speeds up to 16500 rev/min, has resulted in the following conclusions:

1. The maximum temperature of the loaded pad(s) is similar at lower shaft speeds, regardless of whether the load is "on pad" or "between pads". Only at speeds above 12000 rev/min does the difference become noticeable, when LOP is hotter.

2. Reducing the "bearing" clearance from 0.102 mm to 0.076 mm had little effect on the operating temperature of the bearing. However, a further reduction to 0.051 mm resulted in a steep rise in the maximum temperature of the babbitt lining.

3. With a load of 11.1 kN, the bearing operated satisfactorily in reverse rotation at speeds up to 12000 rev/min. The tests indicated that higher speeds, at lower loads, are possible.

4. Calculations have shown that substantial reductions in the power loss of the LEG bearing are possible when the oil inlet temperature is raised. In the example cited in this paper, the savings can be as large as 39 per cent in comparison to the losses of the conventional bearing. This assumes that the maximum temperature of both bearings is similar.

5. The test data has been compared with results from a computer model and correlation has been found to be acceptable. This would seem to confirm that the lower operating temperature of the LEG bearing is the result of reducing hot oil carry over.

ACKNOWLEDGEMENTS

The authors wish to thank the National Research Council of Canada and Kingsbury, Inc. for permission to publish this paper. Special thanks go to Mr. D. Kleinbub, and other staff of the NRC Tribology and Mechanics Laboratory, for their help in carrying out this study.

REFERENCES

1 **Brockwell, K., Kleinbub, D.** and **Dmochowski, W.** Measurement and Calculation of the Dynamic Operating Characteristics of the Five Shoe, Tilting Pad Journal Bearing. *STLE Tribology Transactions*, 1990, **33**, 481-492.

2 **Ettles, C.M.McC.** The Analysis and Performance of Pivoted Pad Journal Bearings Considering Thermal and Elastic Effects. *ASME Journal of Lubrication Technology*, 1980, **102**, 182-192.

3 **Ettles, C.M.McC.** The Analysis of Pivoted Pad Journal Bearings Assemblies Considering Thermoelastic Deformation and Heat Transfer Effects. *STLE Tribology Transactions*, 1992, **35**, 156-162.

4 **Brockwell, K.** and **Dmochowski, W.** Thermal Effect in the Tilting Pad Journal Bearing. *J. Phys. D: Appl. Phys.*, 1992, **25**, 384-392.

5 **Mikula, A. M.** Further Test Results of the Leading-Edge-Groove (LEG) Tilting Pad Thrust Bearing. *ASME Journal of Lubrication Technology*, 1988, **110**, 174-180.

6 **Dmochowski, W., Brockwell, K., DeCamillo, S.** and **Mikula, A.** A Study of the Thermal Characteristics of the Leading Edge Groove And Conventional Tilting Pad Journal Bearings. Submitted to *ASME Journal of Tribology*.

7 **Brockwell, K.** and **Kleinbub, D.** Measurements of the Steady State Operating Characteristics of the Five Shoe, Tilting Pad Journal Bearing. *STLE Tribology Transactions*, 1989, **32**, 267-275.

8 **Constantinescu, V.N.** Basic Relationships in Turbulent Lubrication and Their Extension to Include Thermal Effects. *ASME Journal of Lubrication Technology*, 1973, **94**, 147-154.

9 **Frene, J.** and **Constantinescu, V., N.** Operating Characteristics of Journal Bearings in the Transitional Region. Proceedings of 2nd Leeds-Lyon Symposium on Tribology, *Superlaminar Flow in Bearings*, 1975, paper VII(i), pp. 121-124, (Mechanical Engineering Publication, London)

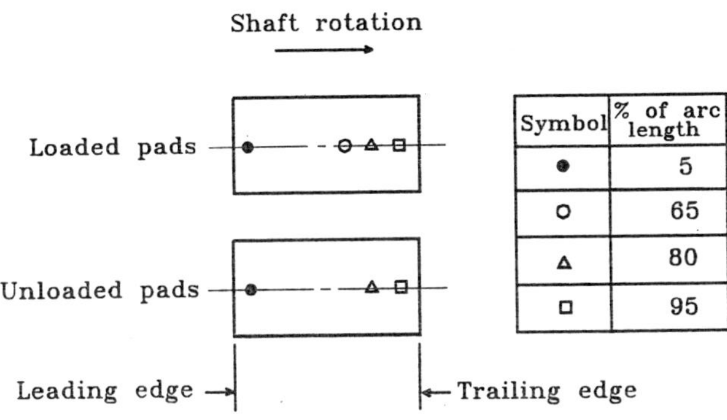

Fig.1 Thermocouple locations

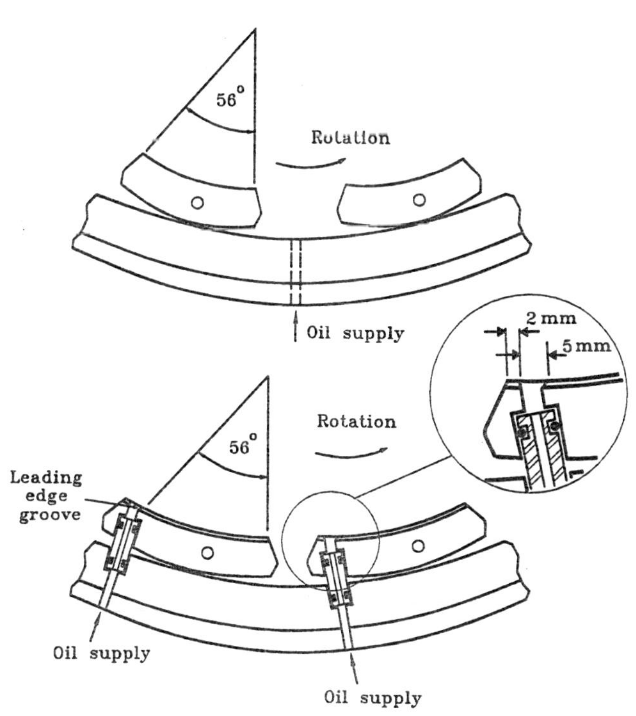

Fig.2 Pad arrangements:
a) conventional   b) LEG

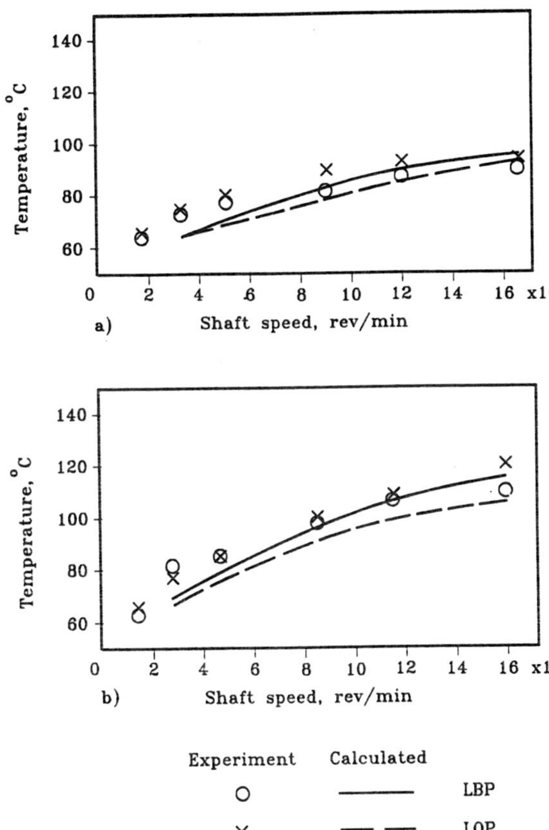

Fig.3 Maximum pad temperatures for LBP and LOP Group 2 bearing, nominal flow rates
a) 5.18 kN (1163 lb) load
b) 11.1 kN (2500 lb) load

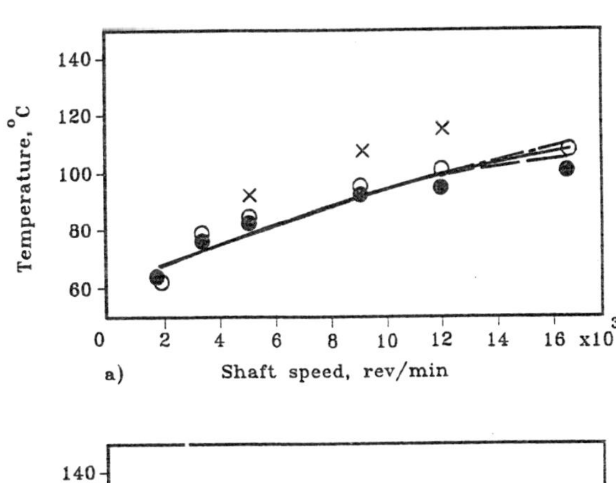

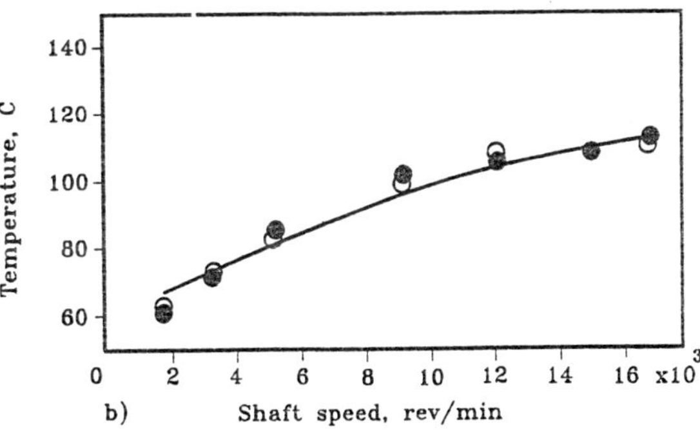

Fig.4 Maximum pad temperatures for different bearing clearances
Nominal flow rates, 11.1 kN (2500 lb) load
a) LBP   b) LOP

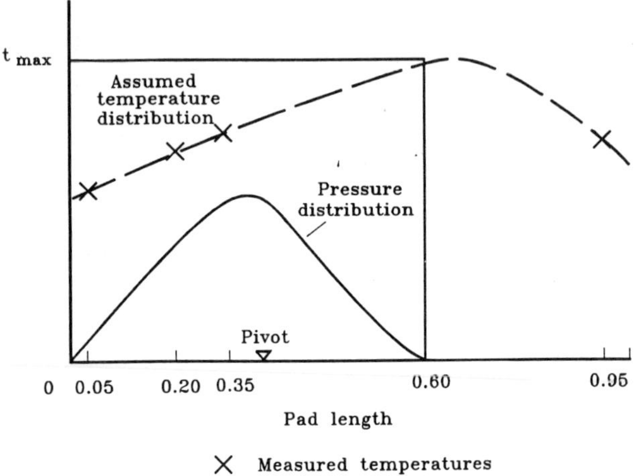

Fig.5 Estimation of maximum pad temperature - reverse rotation tests

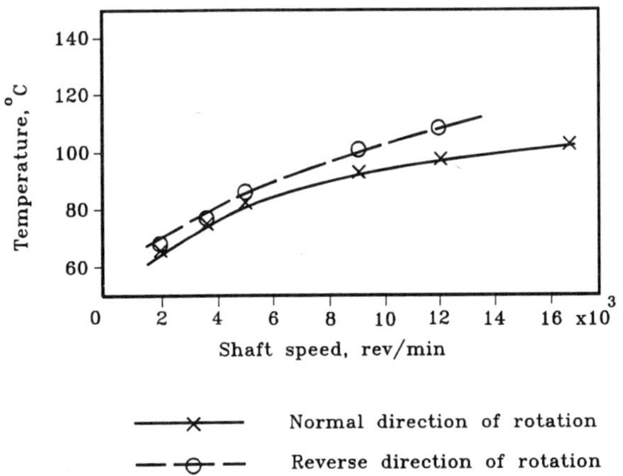

Fig.6 Maximum pad temperatures for forward and reverse directions of rotation
Group 1 LEG bearing, nominal flow rates, 11.1 kN (2500 lb) load, LBP

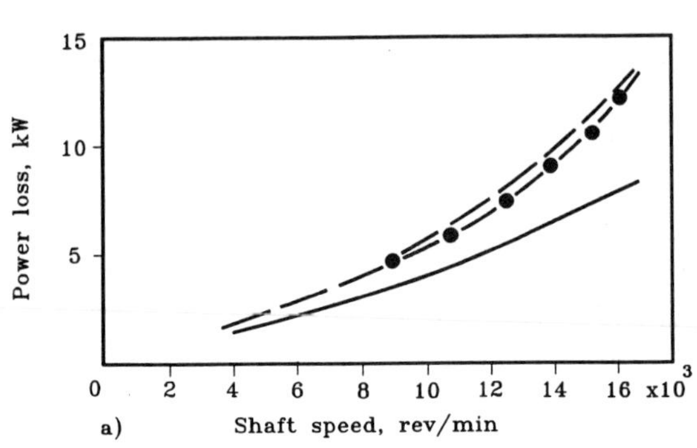

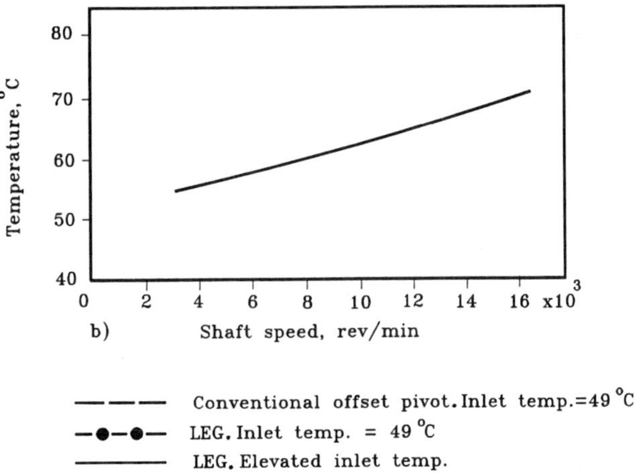

Fig.7 Power loss analysis
Group 1 bearings, nominal flow rates, 11.1 kN (2500 lb) load, LBP
a) calculated power losses
b) elevated oil supply temperatures - LEG bearing

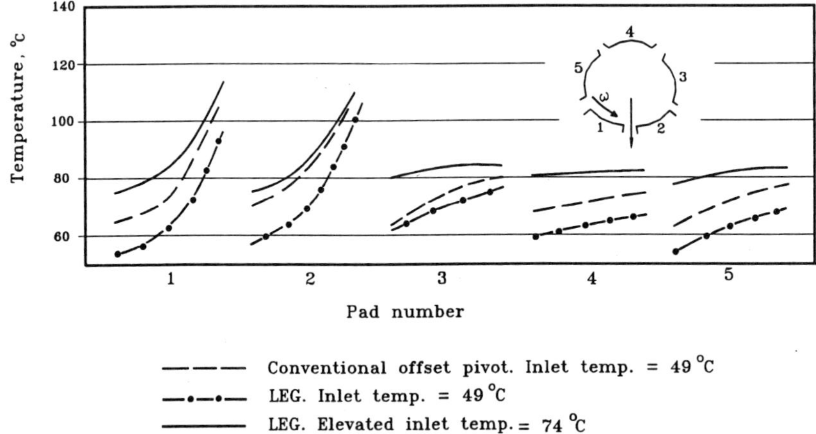

Fig.8 Calculated pad temperature profiles for 16500 rev/min
Group 1 bearings, nominal flow rates, 11.1 kN (2500 lb) load, LBP

Frictional losses in a spherical hydrodynamic journal bearing

I A CRAIGHEAD, BSc, PhD, MIMechE
Division of Dynamics and Control, University of Strathclyde, Glasgow, UK
P S LEUNG, BSc, MSc, PhD
Department of Mechanical Engineering, University of Northumbria, Newcastle upon Tyne, UK

SYNOPSIS   In many machines it is necessary to employ one or more cylindrical journal bearings to carry the lateral load and a thrust bearing to carry axial loads.   The use of a journal bearing with a spherical geometry has the potential to fulfil the requirements of both a journal and a thrust bearing simultaneously.   The spherical journal bearing is studied and its operating parameters determined in the presence of an axial load.   Frictional losses are determined for the bearing and then compared to those predicted for a conventional cylindrical journal bearing in conjunction with a thrust bearing under the same duty.   The spherical bearing was found to offer significant reductions in frictional losses.

NOTATION

A_f = turbulent friction/laminar friction

b = thrust bearing pad width

C = bearing radial clearance

D = bearing diameter

E = effective eccentricity ratio for each cylindrical ring

E_o = eccentricity ratio when no axial loading present

f = friction factor

F_j = viscous frictional force

F_R = bearing force along line of centres

$\overline{F}_R$ = non-dimensional bearing force = $\dfrac{1}{\eta(\omega_1+\omega_2)LR}\left(\dfrac{C}{R}\right)\left(\dfrac{L}{R}\right)F_R$

F_s = bearing force acting normal to bearing centres

$\overline{F}_s$ = nondimensional bearing force = $\overline{F}_R F_s / F_R$

F_z = axial force

G_θ = turbulent coefficient

G_β = turbulent coefficient

h = local film thickness

$h*$ = film thickness at start of film rupture

H = power loss

H_z = power loss in thrust bearing

L = bearing length

M_j = viscous moment on journal

$\overline{M}_j$ = non-dimensional viscous moment = $\dfrac{C}{\eta R^4(\omega_1+\omega_2)}M_j$

N = rotational speed (rev/sec)

p – fluid film pressure

R = sphere radius

Re = local Reynolds number

R_{eb} = bearing Reynolds number = $\dfrac{\pi\rho D(N_1-N_2)C}{\eta}$

R_{em} = mean Reynolds number for each cylindrical ring

S = Sommerfeld number = $\dfrac{\eta(N_1+N_2)LD}{W}\left(\dfrac{R}{C}\right)^2$

W = resultant bearing load

Z = axial coordinate

z = axial displacement

$\overline{z}$ = non-dimensional axial displacement z/C

α = ratio of film thickness

β = polar coordinate

βe = max value of β

$\gamma(\theta)$ = Sommerfeld variable

ϵ = bearing eccentricity ratio

θ_n = angles associated with film rupture and reformation

η = fluid viscosity

ρ = fluid density

τ_{JC} = viscous shear stress in cavitation region

τ_{JF} = viscous shear stress in full film region

ϕ = bearing attitude angle

ω_1 = rotational speed of journal (rad/sec)

ω_2 = rotational speed of bush (rad/sec)

1   INTRODUCTION

Advances in CNC machines over the last decade have enabled the manufacture of accurate spherical surfaces to be achieved almost as easily and economically as cylindrical surfaces.   This opens up the possibility of using spherical journal bearings in place of conventional cylindrical bearings for virtually any application.   The main advantage that a spherical bearing has over a cylindrical bearing is its ability to resist axial loads in conjunction with lateral loads.   For certain operational ranges a spherical bearing can replace a conventional journal bearing and a thrust bearing thus reducing the cost of the machine and reducing the space devoted to housing the bearings.   An added advantage of the spherical bearing is its insensitivity to shaft misalignment.   Large angular misalignment will have little effect on the performance of the bearing unlike conventional journal bearings and thrust bearings (1,2).

Spherical bearings have been used for a number of years (3,4,5) but were usually restricted to specialised applications because of the difficulty and high cost of manufacture.   Analyses of this type of bearing were usually confined to the specialised applications under consideration (6,7,8) and it was not until 1980 that Goenka & Booker (9) presented a more general analysis based on the finite element method.   Leung et al (10) adapted the finite difference approach to take account of the spherical geometry.   Their analysis was restricted to situations where there was no axial load present.   The results showed that there was very little difference between a conventional cylindrical journal bearing and an equivalent spherical journal bearing.   This analysis has been extended to account for axial loading in the present work.

2 THE REYNOLDS EQUATION FOR A SPHERICAL BEARING

Figure 1 shows the geometry of a spherical bearing under the action of an axial load. The Reynolds equation for such a bearing can be stated as:

$$\frac{1}{R^2\cos^2\beta}\frac{\partial}{\partial\theta}\left(G_\theta\frac{h^3}{\eta}\frac{\partial p}{\partial\theta}\right)+\frac{1}{R^2}\frac{\partial}{\partial\beta}\left(G_\beta\frac{h^3}{\eta}\frac{\partial p}{\partial\beta}\right)+\frac{1}{R^2}\theta\tan\beta\frac{\partial}{\partial\theta}\left(G_\beta\frac{h^3}{\eta}\frac{\partial p}{\partial\beta}\right)$$

$$+\frac{1}{R^2\beta}\frac{\partial}{\partial\beta}\left(G_\beta\frac{h^3}{\eta}\theta\tan\beta\frac{\partial p}{\partial\theta}\right)+\frac{1}{R^2}\theta\tan\beta\frac{\partial}{\partial\theta}\left(G_\beta\frac{h^3}{\eta}\theta\tan\beta\frac{\partial p}{\partial\theta}\right)-\frac{(\omega_1+\omega_2)}{2}\frac{\partial h}{\partial\theta}+\frac{\partial h}{\partial E} \tag{1}$$

where G_θ and G_β are coefficients dependent on the local Reynolds number to take account of superlaminar flow (11) and are given by:

$$G_\theta = \frac{1}{12+0.0136\,Re^{0.9}} \tag{2}$$

$$G_\beta = \frac{1}{12+0.0043\,Re^{0.96}}$$

For the spherical geometry the film thickness becomes a function of ϵ, θ, β and z is given by:

$$h = C(1+\epsilon\cos\theta\cos\beta - z\tan\beta) \tag{3}$$

Solution of equation (1) can be achieved by finite difference methods (12) but it requires a large amount of cpu time owing to its complexity. In (10) a reduced equation:

$$\frac{1}{R^2\cos^2\beta}\frac{\partial}{\partial\theta}\left(G_\theta\frac{h^3}{\eta}\frac{\partial p}{\partial\theta}\right)+\frac{1}{R^2}\frac{\partial}{\partial\beta}\left(G_\beta\frac{h^3}{\eta}\frac{\partial p}{\partial\beta}\right)$$

$$=\frac{(\omega_1+\omega_2)}{2}\frac{\partial h}{\partial\theta}+\frac{\partial h}{\partial t} \tag{4}$$

was found to be sufficiently accurate for values of βe up to 45°. This simplification was adopted for the present study. The equation was non-dimensionalised by substituting:

$$\bar{h} = h/C$$

$$\bar{p} = \frac{p}{\eta(\omega_1+\omega_2)}\left(\frac{C}{R}\right)^2$$

$$\bar{v} = \frac{1}{C(\omega_1+\omega_2)}\frac{\partial h}{\partial t}$$

Solution of equation (4) was carried out for a specified eccentricity ratio and results in the pressure distribution for the bearing. The bearing forces acting along and normal to the line joining the sphere centres can then be obtained by integrating the bearing pressure field:

$$F_R = -\int_{-\beta e}^{\beta e}\int_0^{2\pi} p\,R^2\cos^2\beta\sin\theta\,d\theta\,dB$$

$$F_s = \int_{-\beta e}^{\beta e}\int_0^{2\pi} p\,R^2\cos^2\beta\sin\theta\,d\theta\,dB \tag{5}$$

And also

$$W = (F_R^2 + + F_s^2)^{\frac{1}{2}}$$

The attitude angle is given by

$$\phi = \tan(F_s/F_R) \tag{6}$$

and the Sommerfeld number is derived from:

$$S = \frac{1}{\pi\left(\bar{F}_{R}^2+\bar{F}_s^2\right)^{1/2}}\left(\frac{L}{R}\right) \tag{7}$$

If an axial load is then applied to a bearing operating at a given eccentricity ratio and corresponding attitude angle and Sommefeld number a new position of equilibrium must be achieved. This was found by iterating values of ϵ and z according to:

$$\epsilon_{new} = \epsilon_{old}(1+0.5(W_{old}-W_{new})/W_{old})$$

$$z_{new} = Z_{old}(1+0.25(F_{zold}-F_{znew})/F_{zold}) \tag{8}$$

until the original Sommerfeld number was achieved in the presence of the required axial force. This scheme operated satisfactorily over the range of eccentricity ratios 0.1–0.8. The change in pressure field is shown in Figure 2 as the axial force is increased for an initial eccentricity ratio of 0.7. The maximum axial displacement that the journal can tolerate before contact occurs between shaft and bearing is given by:

$$Z_{max} = C(1-\epsilon\cos\beta_e)/\tan\beta_e \tag{9}$$

However, it is to be expected that as the surfaces approached each other, lubricant pressures would reach very high values and also as it would be inadvisable to routinely operate bearings in this condition so a practical maximum axial displacement was considered to be when the two surfaces came within 5% of C of each other. This meant that the performance of the bearing at $\epsilon = 0.9$ could not be obtained but it was decided that this would be of limited interest.

It was found that the eccentricity ratio and attitude angle changed with axial loading and these variations are shown in Figures 3 and 4 for bearings of $L/D = 0.5$ $(\beta_e = 30°)$ and $L/D = 0.707$ $(\beta_e = 45°)$. The variation in axial loading with axial displacement was found to depend on the angle β_e for the bearing. Figure 5 shows this relationship for the two cases $(\beta_e = 30°$ and $45°)$.

3 FRICTIONAL POWER LOSS IN THE SPHERICAL BEARING

The frictional power loss in the journal bearing is dependent on the viscous shear force of the lubricant which in turn is governed by the relative velocity between the journal and the bearing surfaces. For a given rotational speed of the journal ω_1 and of the bearing bush ω_2 the local surface stresses vary both with θ and β for a spherical bearing. The spherical bearing can be considered to consist of a series of cylindrical rings with varying diameters and infinitely small width. Based on the Navier–Stokes equation, and with the presence of a fully developed film, the local viscous shear stress for each of these cylindrical rings can be approximated by:

$$\tau_{JF} = A_f\left[\frac{1}{2R\cos\beta}\frac{\partial p}{\partial\theta}h + \frac{\eta}{h}(\omega_1-\omega_2)R\cos\beta\right] \tag{10}$$

A_f is the ratio of turbulent viscous friction to laminar viscous friction and for this study the work of Smith & Fuller (13) was adopted where:

$$A_f = 0.039\,Re_m^{0.57} \quad \text{for turbulent flow}$$

$$A_f = 1 \quad \text{for laminar flow} \qquad (11)$$

Equations 11 do not provide a continuous representation of A_f in the transition region so for the values of R_{em} between 1000 and 1500 a linear interpolation was used to bridge between the two types of flow.

The presence of an axial displacement z affects the fricitional losses in several ways. The pressure field will change due to the reduction in h and also R_{em} for each cylindrical ring will be modified as the clearances change with displacement in the axial direction.

For the cavitation region, it is assumed that the lubricant circumferential flow appears as narrow fingers and the proportion of the wetted area is a fraction of the corresponding area in the full film region. The condition that $dp/d\theta = p = 0$ in the cavitation region was also adopted in this study. This allows the viscous shear stress in this region to be given by:

$$\tau_{Jc} = \alpha\,A_f\,\frac{\eta}{h}\,(\omega_1 - \omega_2)\,R\cos\beta \qquad (12)$$

and from the continuity of flow (14) α is equal to $\frac{h^*}{h}$.

The viscous frictional force and hence moment acting on the journal can then be determined by integrating the local shear stress with respect to θ and B:

$$F_J = \int_{-\beta e}^{\beta e}\int_0^{2\pi} \tau_J\,R^2\cos\beta\,d\theta\,d\beta \qquad (13)$$

$$M_J = \int_{-\beta e}^{\beta e}\int_0^{2\pi} \tau_J\,R^3\cos^2\beta\,d\theta\,d\beta \qquad (14)$$

In a journal bearing it is more usual to express the frictional losses as a power loss which is defined as:

$$\text{Power} = M_J\,\omega_1 \qquad (15)$$

By substituting (10) and (12) into (14), the non-dimensional viscous moment $\overline{M}_J$ can be obtained from

$$\overline{M}_J = \frac{\epsilon}{2}\int_{-\beta e}^{\beta e}\int_0^{2\pi} Af\overline{p}\cos^2\beta\sin\theta\,d\theta\,d\beta \qquad (16)$$

$$+\lambda\int_{-\beta e}^{\beta e}\frac{Af\cos^3\beta}{(1-E^2)^{1/2}}[\gamma(\theta_2)-\gamma(\theta_1)+\gamma(\theta_4)-\gamma(\theta_3)]\,d\beta$$

$$+\lambda\int_{-\beta e}^{\beta e}\frac{Ah^*\cos^3\beta}{(1-E^2)^{1.5}}[\gamma(\theta_3)-\gamma(\theta_2)-(E\sin\gamma(\theta_3)-E\sin\gamma(\theta_2))]\,d\beta$$

where

$$\lambda = (1-(\omega_2/\omega_1))/(1+(\omega_2/\omega_1))$$

$$\gamma(\theta) = \cos^{-1}\frac{(E+\cos\theta)}{(1+E\cos\theta)} \qquad (17)$$

and $\;E = \epsilon\cos\beta/(1-\overline{z}\sin\beta)$

It is common practice in journal bearing design to express the power loss by means of a friction factor f such that:

$$\text{Power} = f\;W\;R\;\omega_1 \qquad (18)$$

and the friction factor is related to the viscous moment by

$$f\left(\frac{R}{C}\right) = \pi\,S\,\overline{M}_J\left(\frac{R}{L}\right) \qquad (19)$$

Figure 6 shows the friction factor for a spherical bearing $L/D = 0.5$ operating in the laminar regime and also for the turbulent case when $Re_b = 3000$ and 6000. The effect of axial loading is shown when the maximum axial load is present ($10\%\,W$). Figure 7 shows the corresponding results for a bearing of $L/D = 0.707$.

4 FRICTIONAL POWER LOSS OF AN EQUIVALENT CYLINDRICAL JOURNAL PLUS A TILTING PAD THRUST BEARING

From previous work (10) it was found that the frictional losses were almost identical in the spherical bearing and an equivalent cylindrical bearing in the absence of an axial load. The figures agreed closely with those predicted by using ESDU 84031 (15) in which the power loss is given by:

$$H' = \frac{H}{\eta_e N^2 D^2 L}\left(\frac{C_d}{D}\right) \qquad (20)$$

where H' = the non-dimensional power loss obtained from figures 7 or 8 of (15)
H = the power loss
D = journal diameter
L = bearing length
C_d = diametral clearance
N = journal speed rev/sec
η_e = effective dynamic viscosity

The friction factor is related to the non-dimensional power loss as follows:

$$f\left(\frac{R}{C}\right) = \frac{H'S}{\pi} \qquad (21)$$

As a cylindrical bearing can sustain no axial load, a thrust bearing is often required when axial loads are present. Additional frictional losses will arise due to the action of the thrust bearing and these must be added to the journal losses to enable a comparison to be made with the losses in a spherical bearing. Reference (16) gives guidelines for the determination of frictional losses in steadily loaded, offset-pivot, tilting pad thrust bearings. The power loss is given by equation 6.5 (16) as:

$$H_z = \frac{30\,h\,N\,d_m\,F_z}{b} \qquad (22)$$

where h = minimum film thickness
N = rotational speed
d_m = mean pad diameter
F_z = axial load
b = pad width

Non–dimensionalising the thrust bearing loss by dividing by the journal bearing power loss results in:

$$\frac{H_z}{H'\eta_e N^2 D^2 L}\left(\frac{C}{R}\right) = H'_z = \frac{30 h N d_m F_z}{b H'\eta_e N^2 D^2 L}\left(\frac{C}{R}\right) \qquad (23)$$

therefore

$$H'_z = \frac{30 h d_m F_z}{H' L D \eta_e N L D}\left(\frac{C}{R}\right)^2 \left(\frac{R}{C}\right)\left(\frac{F_z}{W}\right)$$

$$H'_z = \frac{30}{SH'}\left(\frac{h}{b}\right)\left(\frac{d_m}{D}\right)\left(\frac{R}{C}\right)\left(\frac{F_z}{W}\right) \qquad (24)$$

There are many possible thrust bearing configurations so for comparison purposes typical values of the non–dimensional parameters on the right hand side of equation 24 were taken as $h/b = C/R$, $dm/D = 1.3$ which resulted in

$$H'_z = \frac{39}{SH'}\left(\frac{F_z}{W}\right) \qquad (25)$$

Using equation 25 and the appropriate values of S and H' (from (15)) values of H'_z were determined as a percentage increase in cylindrical journal bearing losses. Reference (16) goes on to suggest that churning losses could be significant and these are given tentatively as:

$$H_{ch} = K \rho N^3 D_R^5 \left(1 + \frac{4t}{D}\right) \qquad (26)$$

where K = fn of Reynolds No $(\rho N D_R^2/\eta)$
    D_R = Diameter of runner
    t runner thickness

Once again the actual value is dependent on the bearing configuration but typical figures result in an increase in thrust bearing losses of up to 50%. Bearing in mind that the determination of churning losses is tentative and that they can be reduced by directed lubrication a value of 25% increase was added to the calculated thrust bearing losses.

The total losses of the equivalent cylindrical journal bearing plus thrust bearing were estimated on these lines and then converted to a friction factor. Their comparison with the losses of a spherical journal bearing under the same duty is shown in Figures 8 and 9.

5 DISCUSSION

The axial load capabilities of the spherical bearing are illustrated in Figure 5. With $\beta_e = 30°$, axial loads of up to 10% of the lateral load can be carried whereas for $\beta_e = 45°$ the axial load can be increased to 25%. It is envisaged that larger values of β_e would enable the bearing to withstand higher axial loads. Investigation of such bearings would require the use of equation 1 rather than 4. However, one drawback of a bearing with $\beta_e > 45°$ would be the relatively small shaft diameter compared to the bearing diameter. Bearings of this configuration would probably only be contemplated for specialised applications.

The friction factors predicted for the spherical bearings in Figures 6 and 7 were somewhat unexpected. It can be observed that for most cases the application of an axial load results in a reduction in friction factor rather than an increase as might be envisaged. This is believed to be due to the various influences that a z displacement

of the journal produces. The most noticeable change that occurs is an increase in the peak pressure within the bearing (Figure 2). It was thought that the reduction in h corresponding to the higher pressure would increase the viscous drag but careful consideration reveals that the peak pressure acts over a reduced area of the bearing and also that it will occur at a smaller radius than for the situation with $F_z = 0$. The second influence that a z displacement introduces is the modification to the local Reynolds number within the bearing. For a positive z displacement the film thickness is reduced for positive values of Z and increased for negative values of Z. This in turn will modify the local Reynolds number for each cylindrical strip within the model and increase the loss due to turbulence for half of the bearing ($-ve$ Z values) and reduce the loss due to turbulence for the other half ($+ve$ Z values). The two changes will not be equal due to the non–linear relationship assumed for the shear stress in equation 10. It can be seen from Figures 6 and 7 that the axial loading increases the friction factor for the turbulent bearings suggesting that the increase in turbulence due to a z displacement exceeds the other influences on friction factor that arise.

When compared to a conventional cylindrical journal bearing and thrust bearing operating under the same conditions it can be seen from Figures 8 and 9 that the spherical bearing shows substantial reductions in friction factor. For a bearing with $L/D = 0.5$ the friction factor can be reduced to 50% whereas for a bearing with $L/D = 0.707$ the friction factor can be reduced to as low as 35%. The greatest reductions occur when the bearing is operating in the laminar regime which covers most small and medium sized bearings operating at all but the highest speeds. For larger and faster bearings operating in the superlaminar regime the use of the spherical geometry has been considered for a floating ring bearing and shown considerable reductions in frictional losses compared to a conventional journal bearing (18).

Although the figures quoted for the cylindrical journal and thrust bearings are generalised, the method outlined could be used to determine likely savings in specific cases. The typical improvements indicated by this study suggest that the application of a spherical journal bearing is worthy of consideration if frictional losses are to be reduced in a machine which possesses both journal bearings and a thrust bearing.

6 CONCLUSIONS

A hydrodynamic journal bearing with a spherical geometry has been analysed. It was found that axial loads up to 25% of the lateral load could be carried with no significant increase in frictional losses. Comparison of the spherical bearing with a cylindrical journal bearing and a tilting pad thrust bearing operating under the same conditions indicated that frictional losses for the spherical bearing could be as low as 35% of the losses for the conventional arrangement.

REFERENCES

(1) CRAIGHEAD, I.A. et al. The influence of thermal effects and shaft misalignment on the dynamic behaviour of fluid–film bearings. Procs 6th Leeds–Lyons Symp Trib, Lyons, 1979, 47–55.

(2) STANGO, R.J. & JUNGMANN, R.H. A variational method for evaluating thrust bearing element load distribution. *ASME Journal for Industry*, 1989, 79–86.

(3) SHAW, M.C. & STRANG, C.D.   The hydrosphere – a new hydrodynamic bearing. *Journal Applied Mechanics*, 1948, <u>70</u>, 137–145.

(4) GROSS, G.A.   Fluid film lubrication.   Publ. Wiley 1980.

(5) SAUTTER, S.   Spherical plain bearings in segment gates – an economic solution. *Ball bearing journal*, <u>230</u>.

(6) WANNIER, G.H.   A contribution to the hydrodynamics of lubrication. *Quarterly Applied Mathematics,* 1950, <u>8</u>, 1–32.

(7) PAN, C.H.T.   Gas lubricated spherical bearings. *ASME Journal of Basic Engineering*, 1968, <u>85</u>, 311–323.

(8) GOENKA, P.K.   Effect of surface ellipticity on dynamically loaded spherical and cylindrical joints and bearings.   PhD Thesis, 1980, Cornell University.

(9) GOENKA, P.K. & BOOKER, J.F.   Spherical bearings: static and dynamic analysis via the finite element method. *ASME Journal of Lubrication*, <u>102</u>, 308–319.

(10) LEUNG, P.S. et al.   An analysis of the steady state and dynamic characteristics of a spherical hydrodynamic journal bearing. *ASME Journal of Tribology*, 1989, <u>111</u>, 459–467.

(11) CONSTANTINESCU, V.N.   Basic relationships in turbulent lubrication and their extension to include thermal effects.   Trans ASME paper 72–lub–16.

(12) LEUNG, P.S.   An investigation of the dynamic behaviour of floating ring bearing systems and their application to turbogenerators.   PhD Thesis, 1988, Newcastle Polytechnic.

(13) SMITH, M.I. & FULLER, D.   Journal bearing operation at superlaminar speeds.   Trans ASME 1956 <u>78</u>.

(14) RUDDY, A.V.   The dynamics of rotor bearing systems with particular reference to the influence of fluid film journal bearings and the modelling of flexible rotors.   PhD Thesis, 1979, University of Leeds.

(15) ESDU Tribology 84031 Calculation methods for steadily loaded axial groove journal bearings, 1984.

(16) ESDU Tribology 83004 Calculation methods for steadily loaded offset pivot, tilting pad thrust bearings, 1983.

(17) ESDU Tribology 85028 Calculation methods for steadily loaded axial groove hydrodynamic journal bearings, superlaminar operation, 1986.

(18) LEUNG, P.S. et al.   An analysis of the steady state and dynamic characteristics of a cylindrical–spherical floating ring bearing.   Procs of IMechE Conf Vib in Rotating Machinery, Heriot–Watt University, 1988.

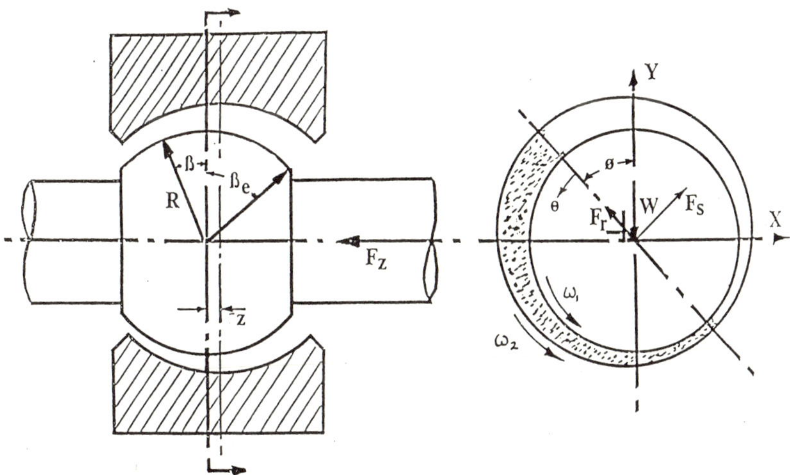

Fig 1   The spherical bearing geometry

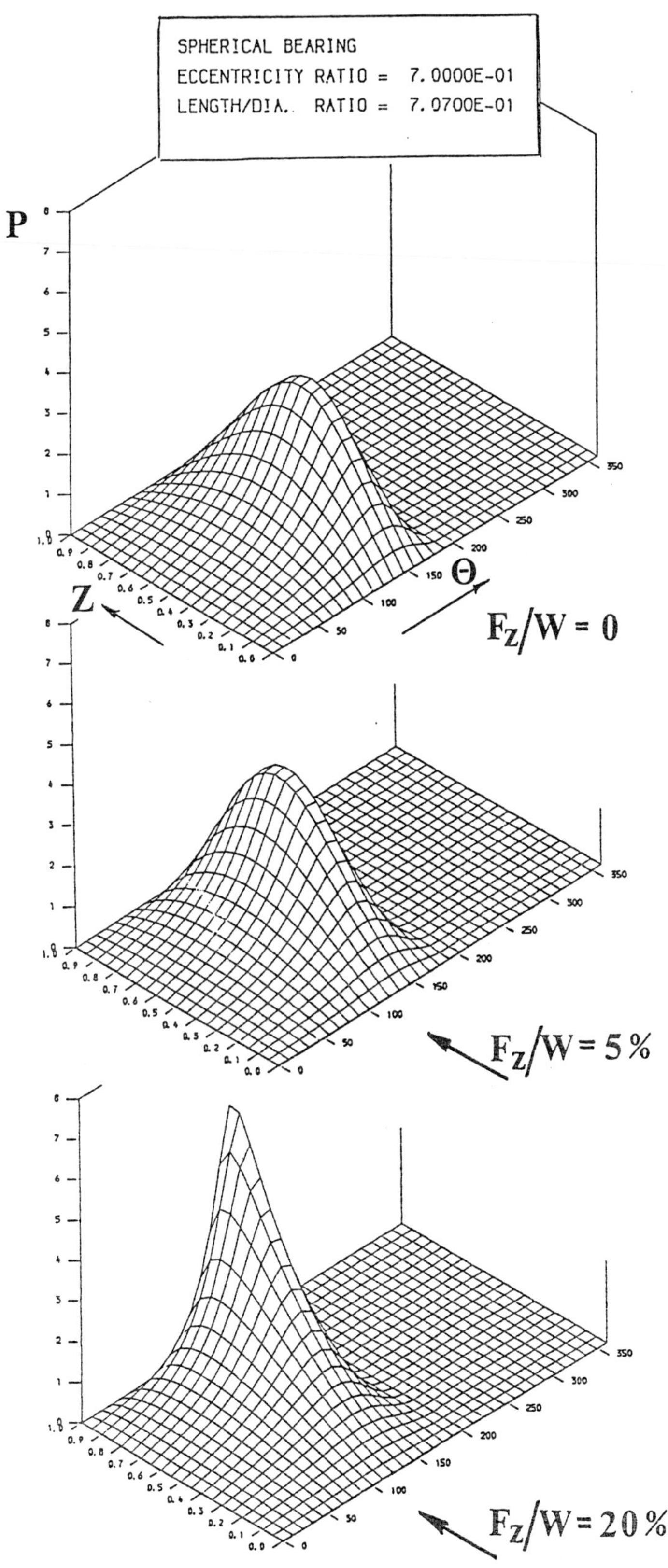

Fig 2   Pressure fields for the spherical bearing
with a lateral and axial load acting

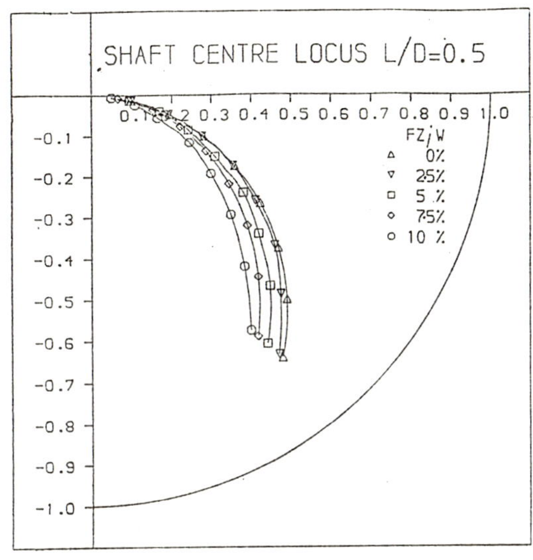

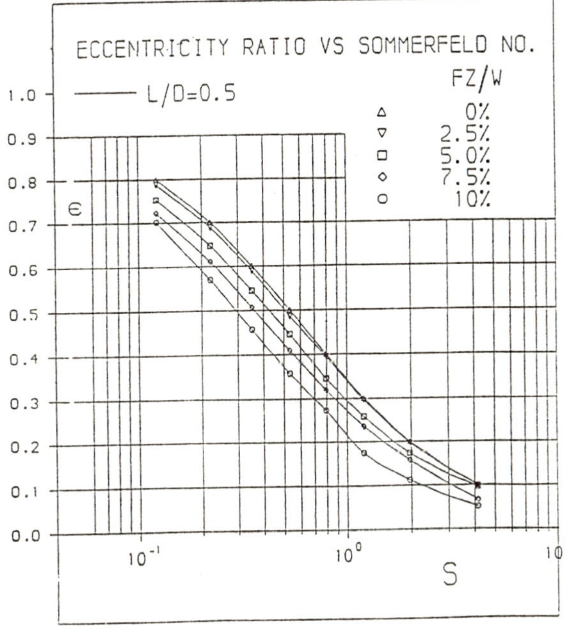

Fig 3   Change in eccentricity ratio and attitude angle with axial load (L/D = 0.5)

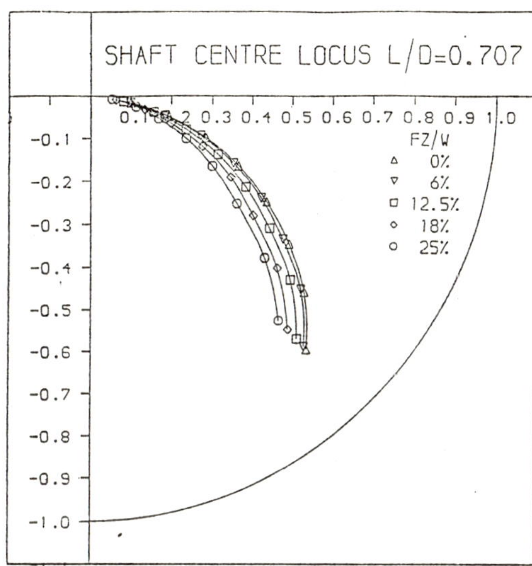

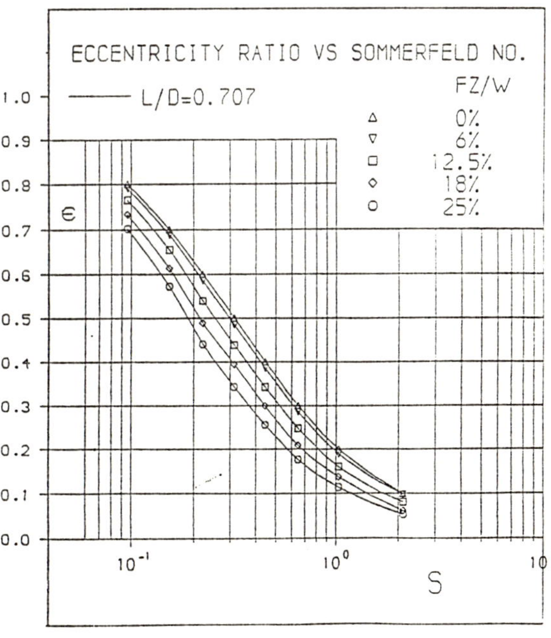

Fig 4   Change in eccentricity ratio and attitude angle with axial load (L/D = 0.707)

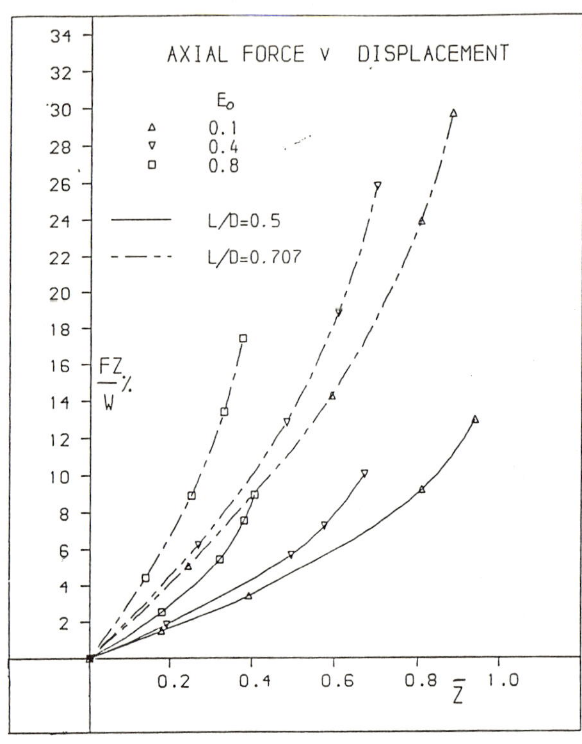

Fig 5   The relationship between axial force and
axial displacement

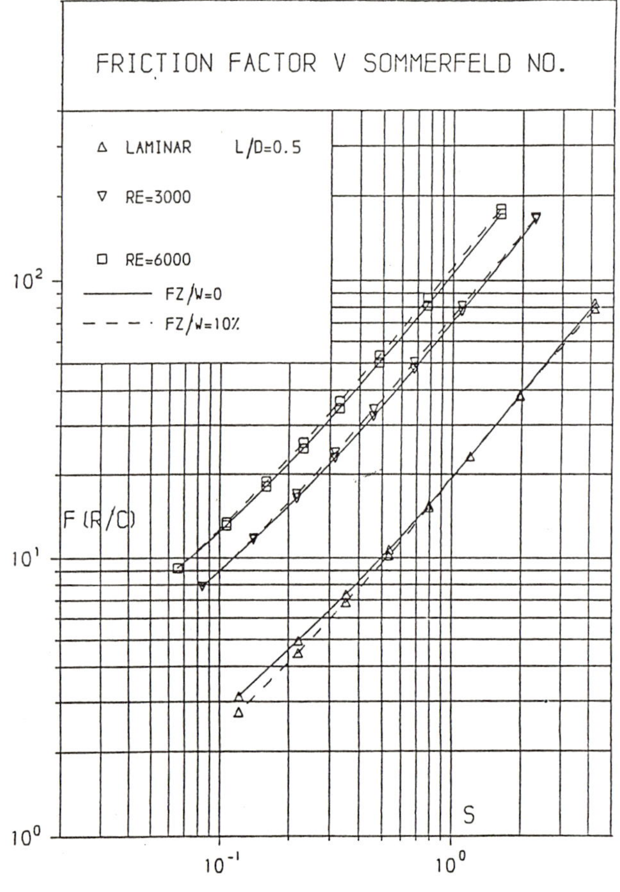

Fig 6   Friction factors for the spherical bearing
(L/D = 0.5)

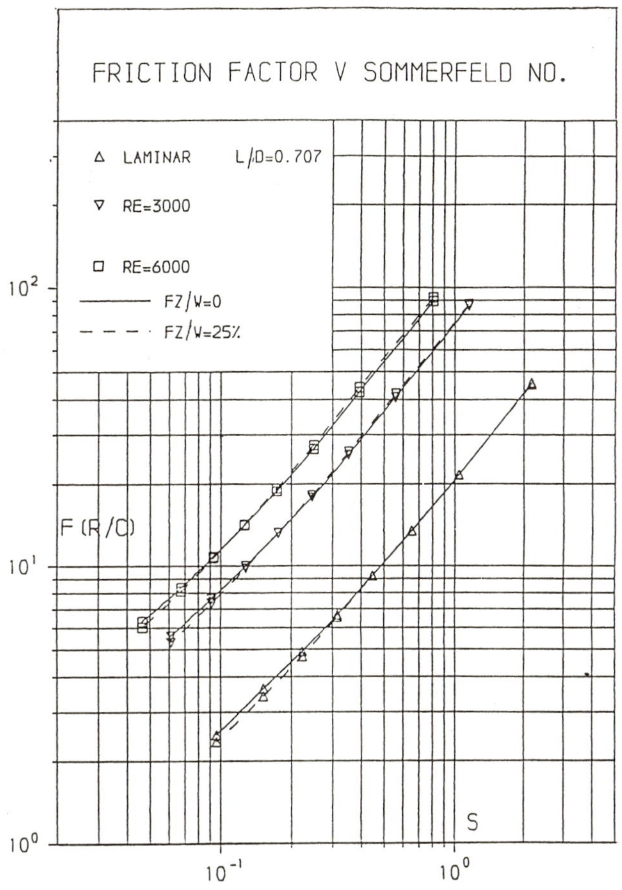

Fig 7   Friction factors for the spherical bearing
(L/D = 0.707)

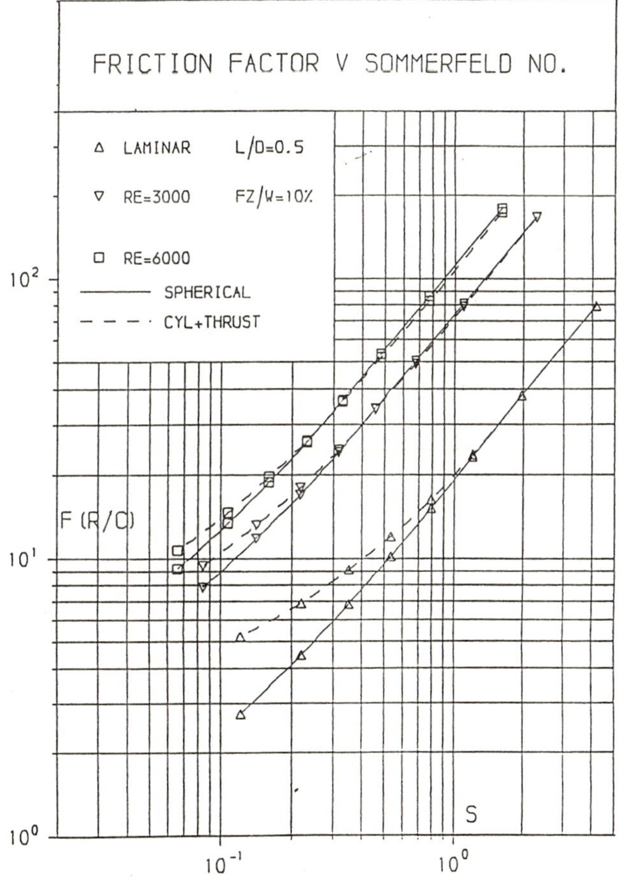

Fig 8   Comparison of friction factors for a spherical
bearing and an equivalent cylindrical bearing
plus thrust bearing (L/D = 0.5)

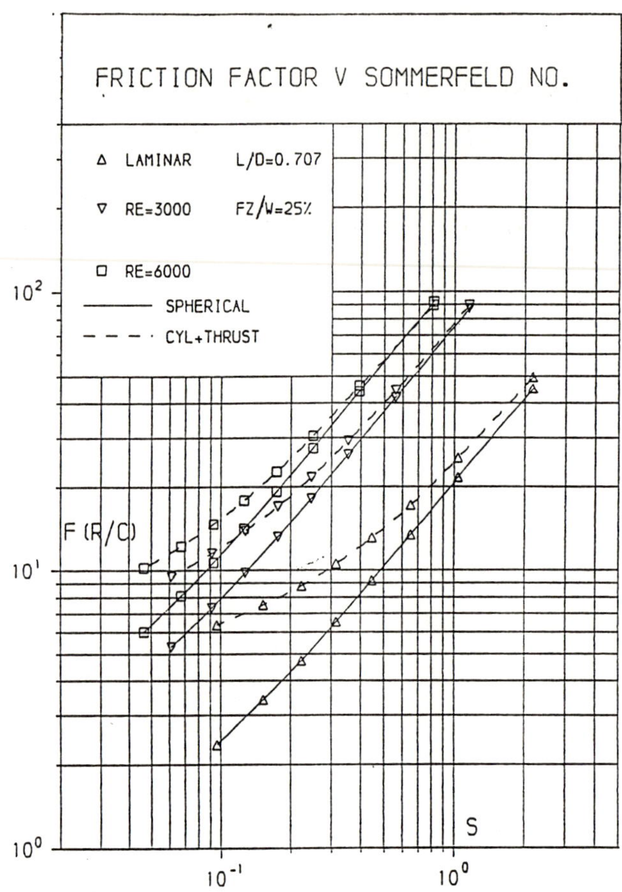

Fig 9 Comparison of friction factors for a spherical bearing and an equivalent cylindrical bearing plus thrust bearing (*L/D* = 0.707)

The effect of increasing fuel costs on the optimum design of a large turbine generator journal bearing

T S WILKINSON, BSc, FIMechE, FIMA, and M SMITH, BSc
NEI Parsons Limited, Newcastle upon Tyne, UK
A C LONGBOTTOM
Loughborough University of Technology, Leicestershire, UK

SYNOPSIS The design/optimisation of a nominally 'elliptical' bearing is described. Thus, an objective function is minimised which includes contributions due to energy dissipation (fuel costs), vibration, stability and drain temperature (penalty costs) and rotor stiffness and oil systems (capital costs). Paradoxically, the optimum design with low and with increased fuel costs is circular.

1   INTRODUCTION

The designers of journal bearings which may be used to support the rotors of large turbine generators must try to satisfy many, sometimes conflicting, requirements. The total energy dissipation rate is only one of them.

Most large turbine generators rotate at 3000 rpm (50 Hz) or in North America at 3600 rpm (60 Hz), however, the rotors must also be supported adequately at all lower speeds (including zero) and, occasionally, at speeds up to 20% higher than normal.

Designers, therefore, must provide hydrostatic support at very low speeds and the cost of such provision is affected by, for example, the nominal pressure based on load and projected area. The cost of the jacking oil system depends on the required pressure and on the oil flow.

Similarly, the cost of the main oil pumps, tanks, pipes, filters, coolers etc. also depends on the (much larger) low pressure oil flow.

The rotors which the bearings support must be flexurally stable, that is, they should not vibrate asynchronously under any operating condition. Furthermore, the response to synchronous forces caused, for example, by imperfect rotor balance should, preferably, be small.

There are, as might be imagined, other requirements, some of which will be discussed in the paper, however, those mentioned already probably suffice to show that energy dissipation (power loss) is but one of several considerations. In principle, it is possible to assign costs to each of them[1] and thus to define the design problem as one of minimising a function of several variables. This approach will be described, together with means by which it may be accomplished and, in keeping with the theme of the seminar, the effect on the optimum design of increasing energy costs relative to others will be described.

2   BEARING GEOMETRY

For more than thirty years Parsons have used elliptical bearings to support turbine generator rotors, that is to say, the vertical diametral clearance is smaller than that horizontally. This is accomplished by assembling bearing halves with shims of thickness $2e\delta$ in the horizontal joints before boring to yield a diameter of $D + 2\delta$ where D is the journal diameter. When the shims are removed the diametral clearance is of course, 2δ horizontally and $2(1 - e)\delta$ vertically. The ellipticity e is usually about 0.4.

In service a light turbine oil is supplied to deep axial grooves adjacent to each horizontal joint and entry to each half is facilitated by relieving the surface over angles of $\pm \theta_w$ and over about 85% of the bearing width (length). This 'washaway' is produced by displacing the boring centre horizontally by a distance and by reducing the boring radius by $a\delta \cos \theta_w$. The depth of the washaway at a horizontal joint is thus $a\delta (1 - \cos \theta_w)$. The washaway depth and angular extent are usually the same on each side of the bearing. If the journal rotates in the θ positive direction (with θ measured from top dead centre), then the first joint is at $\theta = \pi/2$ and the second is at $\theta = 3\pi/2$. In this paper, the washaways of sides one and two may differ, ie. $a_1 \neq a_2$ and $\theta_{w1} \neq \theta_{w2}$,

If the journal is displaced from the geometrical centre of the bearing $X\delta$ horizontally (towards side 2) and $Y\delta$ vertically (downwards) the dimensionless film thickness in the bottom half will be given by

$$h(\theta,z) = 1 + (Y + e)\cos \theta + X \sin \theta + a_1(\sin \theta - \cos \theta_{w1}) \qquad (1)$$
$$\text{for} \quad \pi/2 \leq \theta \leq \pi/2 + \theta_{w1}$$

where the last (washaway) term is omitted for the narrow bands at the edges of the bearing and for $\pi/2 + \theta_{w1} \leq \theta \leq 3\pi/2 - \theta_{w2}$. If the film is not disrupted by then it will certainly be disrupted by the rapid increase of clearance at $\theta = 3\pi/2 - \theta_{w2}$

The dimensionless film thickness in the top half of the bearing is given by

$$h(\theta,z) = 1 + (Y - e)\cos\theta + X\sin\theta + a_2(-\sin\theta - \cos\theta_{w2}) \qquad (2)$$
$$\text{for} \quad 3\pi/2 \le \theta \le 3\pi/2 + \theta_{w2}$$

In the bottom half of the bearing there are two shallow jacking oil recesses (at $\theta = \pi$) and at the $\frac{1}{4}$ and $\frac{3}{4}$ length positions.

3   PERFORMANCE CALCULATIONS

A modified Reynolds' equation was used to calculate performance. In dimensionless form it can be written in the form

$$\frac{\partial}{\partial\theta}(\lambda_2 h^3 \frac{\partial P}{\partial\theta}) + \frac{\partial}{\partial z}(\lambda_3 h^3 \frac{\partial P}{\partial z}) = \frac{\partial h}{\partial\theta} \qquad (3)$$

where the Poiseuille flow coefficients λ_2, λ_3 depend on the local Reynolds' number.

A finite difference method which includes a regular mesh and a direct solution of the resulting set of sparse, linear algebraic equations was used with negative pressures set to zero. As usual the total force generated in the oil film must be vertical and equal to the applied (gravitational) load. The journal position X,Y which uniquely satisfies these conditions must, of course, be found iteratively.

The pressure distribution may be calculated approximately, by using a lubricant viscosity which is independent of position θ,z . Such methods may be more accurate, however, if the 'constant' viscosity corresponds to that of the lubricant when at the drain temperature. The drain temperature was estimated by calculating the local turbulent energy dissipation rate and integrating over the bearing surface and, finally, dividing by the product of the total oil flow and its specific heat.

4   CALCULATION OF BEARING COEFFICIENTS

The eight coefficients which are normally used to establish the dynamic stability of the rotor system and the linearised response to synchronous forces can be calculated easily when the steady running conditions have been satisfied.

The dimensionless displacement coefficients

$$c_{11} = \frac{\partial F_y}{\partial y}, \quad c_{12} = \frac{\partial F_y}{\partial x}, \quad c_{21} = \frac{\partial F_x}{\partial y}, \quad c_{22} = \frac{\partial F_x}{\partial x}$$

and the velocity coefficients

$$b_{11} = \frac{\partial F_y}{\partial \dot{y}}, \quad b_{12} = \frac{\partial F_y}{\partial \dot{x}}, \quad b_{21} = \frac{\partial F_x}{\partial \dot{y}}, \quad b_{22} = \frac{\partial F_x}{\partial \dot{x}}$$

were obtained by adding four right hand sides to that implied by the steady equation 3 which represent, respectively, perturbations of the journal position and velocity in the vertical (y) and horizontal (x) directions. The four solutions thus generated have non-zero pressures in those parts of the steady solution which is deemed to be inapplicable due to cavitation. In such regions the additional solutions are also set to zero.

The coefficients were made dimensionless by dividing absolute values by the steady bearing load, multiplying by the nominal radial clearance

δ and, in the case of the velocity coefficients, by also multiplying by the steady rotational speed Ω.

5   EXAMPLE OF PERFORMANCE CALCULATION

In the case of a typical bearing, designed more than twenty years ago to carry a load of 26.5 tonnes with a journal diameter of 533 mm and with a length of 355 mm (L/D = 2/3), the clearance ratio $2\delta/D$ was 0.002. Such a bearing has been sketched in Fig. 1, the calculated bottom half pressure distribution is shown in Fig. 2. and the dimensionless bottom half film thickness is shown in Fig. 3.

Other useful data are recorded in Table 1.

<u>TABLE 1   OLD (REFERENCE) DESIGN</u>

Bearing geometry

Diameter D	533 mm
Length L	355 mm
Clearance δ	0.508 mm
Nominal viscosity (60°C)	0.0177 Ns/m^2
Ellipticity e	0.442
Rotor stiffness factor S_F	1.0
Washaway extent θ_{w1}	45°
Washaway depth at inlet 1	3 mm
Washaway extent θ_{w2}	45°
Washaway depth at inlet 2	3 mm

Contributions to objective function

Jacking (area) C_J	10.0
Drain temperature C_T	20.0
Oil flow C_Q	10.0
Rotor stiffness C_S	10.0
Energy loss C_H	10.0
Stability C_L	0.0
Odd vibration C_O	30.0
Even vibration C_E	10.0
Total	100.0

6   ROTOR DYNAMICS

Bearing design certainly affects rotor dynamics but the effect on vibrations of small amplitude may differ qualitatively from that of large amplitudes. If, however, a stable bearing design is chosen, the probability that large amplitude non-synchronous vibration will occur is much reduced. Therefore, only small amplitude vibrations will be considered.

Turbine generator rotor systems may be treated, approximately, as several weakly coupled rotors, each symmetrical about its half length and supported by two similar journal bearings.

In Appendix 1 expressions are derived for the journal speed Ω_L at which a rotor will become unstable and for the amplitudes of synchronous vibration due to unbalance. In both cases a simplified rotor model is considered.

7   COST FUNCTIONS

Designers are obliged to consider all those aspects of a design which affect the lifetime cost or value to a customer. The utility should be maximised and/or the lifetime cost (initial cost plus running costs) should be minimised.

Eight contributions to the total cost of a bearing to support a large LP steam turbine will be considered in this paper. None is real: convenient algebraic expressions have been invented for each one as follows:

1.   The cost of the jacking oil system which must supply high pressure oil at a sufficient rate to maintain an adequate film thickness increases as the nominal pressure increases. The cost will be approximated herein by

$$C_J = 10\exp[(LD)_r/LD - 1]$$

where the subscript r refers to the original (reference) bearing.

2.   There is, currently, a target oil outlet (drain) temperature of 73C (which is significantly lower than that of the reference bearing). This temperature should, preferably, not be exceeded, however, for a given energy loss, the heat exchangers which ultimately cool the oil will have a greater surface area for heat transfer if the oil is significantly cooler than 73C. The cost of departure from 73C will, therefore, be represented by

$$C_T = 20[(T_0 - 73)/(T_{0r} - 73)]^2$$

where T_{0r} is the outlet temperature of the reference bearing. The cost of the high drain temperature of the reference bearing is thus 20 units. This provides a great incentive to reduce it as will be seen (but not enough to reach 73C).

3.   The total oil flow at rated speed determines the capacity and, therefore the cost, of the pumps, filters, tanks and, of course, the piping system. The cost of the oil system will be represented by

$$C_Q = 10\exp[1.39(Q/Q_r - 1)]$$

so the cost associated with the reference flow rate Q_r is 10 units.

4.   In seeking to improve the dynamic behaviour of a rotor/bearing system by optimising the bearing design, it is natural to reconsider the rotor flexibility to see whether such optimisation might be facilitated by adjusting one or more rotor design parameters. The journal diameter will, in any event, be held constant because a decrease would increase steady and dynamic torsional stresses and an increase would increase capital and running costs. A stiffness factor S_F, normally unity, will be used to multiply each of the natural frequencies of the rigidly supported rotor and, as it is difficult (and expensive) to increase stiffness, increases will be penalised as follows:

$$C_S = 10\exp[5(S_F - 1)\ln 2]$$

This expression effectively doubles the cost if the stiffness is increased by 20% and halves it if reduced by 20%.

5.   The importance of the energy dissipation rate or power loss relative to the capital costs (1-4 above) depends on the cost of fuel and the expected working life of the turbine generator. In this paper a low fuel cost $C_{H1} = 10H/H_r$ and, alternatively, a high cost $C_{H2} = 30H/H_r$, will be considered.

6.   The cost penalties which would be incurred if a rotor/bearing system became flexurally unstable at even a little above rated speed would be large indeed. The cost penalty in the present study has, therefore, been arranged to increase rapidly as the calculated stability limit Ω_L approaches Ω_r where Ω_r is the rated speed. Thus

$$C_L = \exp[(1.2\Omega_r - \Omega_L)/2.6]$$

7.   It is not possible to avoid synchronous vibration due to unbalance and it is not possible a priori to predict likely amplitudes. The first (even) resonant frequency is much lower than the rated speed. Consequently, a rotor which has been balanced, approximately, near the resonant frequency should not vibrate excessively at rated speed. However, the second (odd) resonant frequency is much closer to the rated speed and the difficulty (and cost) of achieving an acceptable state of balance is relatively much greater.

The cost of odd vibration will, therefore, be calculated from

$$C_O = 30U_O\delta /U_{Or}\delta_r$$

where U_O is the dimensionless, odd amplitude of vibration at the rated speed.

The cost of even vibration will, however, be one third of the size, ie.

$$C_E = 10U_E\delta /U_{Er}\delta_r$$

8   DESIGN PARAMETERS

Nine parameters will be selected to minimise an objective function calculated by summing the eight costs already described. Seven of the parameters $\delta, L, e, a_1, a_2, \theta_{W1}$ and θ_{W2} are geometrical and unlimited in range. However, only small variations of the other two will be permitted. In the case of the rotor stiffness parameter $0.8 \leq S_F \leq 1.2$. The second non-geometrical parameter is the nominal viscosity of the lubricant: one less and one more viscous grade will be considered.

9   RESULTS AND DISCUSSIONS

It is not easy to follow all of the decisions made by the optimiser. The principal changes concerned the reduction of length L and the reduction of nominal clearance which then became possible. The reduction of ellipticity might be considered to be surprising in view of the widespread belief that ellipticity increases rotor/bearing stability, however, the authors have noted an increase of stability as ellipticity is reduced in a previous study.

The results of the first optimisation, with a low fuel cost, are summarised in Table 2. The pressure distribution in the bottom half of the narrower more heavily loaded bearing is presented in Fig. 4. The dimensionless film thickness in the bottom half is presented in Fig. 5.

TABLE 2   OPTIMISED DESIGN: case 1

Bearing geometry

Diameter D	533 mm
Length L	229 mm
Clearance δ	0.21 mm
Nominal viscosity (60°C)	0.013 Ns/m^2
Ellipticity e	0.0
Rotor stiffness factor S_F	0.975
Washaway extent θ_{W1}	60°
Washaway depth at inlet 1	0.44 mm
Washaway extent θ_{W2}	55°
Washaway depth at inlet 2	0.93 mm

Contributions to objective function

Jacking (area) C_J	17.5
Drain temperature C_T	1.9
Oil flow C_Q	7.3
Rotor stiffness C_S	9.17
Energy loss C_H	5.98
Stability C_L	0.0
Odd vibration C_O	10.74
Even vibration C_E	5.15
Total	57.76

The effects of increasing the fuel costs by a factor of three, relative to the others, are equally difficult to explain. There has been a further reduction of length and of clearance (Table 3). The pressure is distributed as shown in Fig. 6 and the corresponding film thicknesses are as shown in Fig. 7.

The analysis program allows journal tilt to be considered, however, the asymmetries in these Figures is due to the use of an idiosyncratic contour plotting program.

TABLE 3   OPTIMISED DESIGN: case 2

Bearing geometry

Diameter D	533 mm
Length L	216 mm
Clearance δ	0.19 mm
Nominal viscosity (60°C)	0.013 Ns/m^2
Ellipticity e	0.0
Rotor stiffness factor S_F	0.98
Washaway extent θ_{W1}	60°
Washaway depth at inlet 1	0.37 mm
Washaway extent θ_{W2}	60°
Washaway depth at inlet 2	0.87 mm

Contributions to objective function

Jacking (area) C_J	19.33
Drain temperature C_T	1.61
Oil flow C_Q	6.94
Rotor stiffness C_S	9.31
Energy loss C_H	16.9
Stability C_L	0.0
Odd vibration C_O	9.68
Even vibration C_E	4.46
Total	68.24

10   CONCLUSIONS

In the case of turbine generator bearings, the cost of energy dissipation is only one of many design considerations.

Design is or should be, a matter of maximising utility or minimising lifetime costs. When capitalising energy dissipation costs it may be necessary to estimate fuel costs many years hence.

Seven of the many geometrical parameters which together define a class of elliptical bearings and two non-geometrical parameters have been varied in minimising an objective function.

Compared with a bearing designed for a similar duty more than twenty years ago, the optimised design is of a shorter (more heavily loaded) bearing with a smaller clearance. The washaway depths have been reduced but their angular extent has been increased. A lighter grade of oil has also been substituted.

The effect on the optimum design of increasing fuel costs by a factor of three has been calculated. The bearing length and nominal clearance were reduced further.

Paradoxically, one of the geometrical parameters was the ellipticity but the optimum designs were nominally circular.

REFERENCES

(1)   LONGBOTTOM, A.C.   Optimisation of journal bearing design using coarse grained parallelism.   M.Sc. project report. Loughborough University of Technology, 1990.

(2)   WILKINSON, T.S.   Flexural vibration of turbine generators: simple models of symmetrical rotors.   ASME Conference on vibration and noise, Boston, 1987.

ACKNOWLEDGEMENTS

The authors thank the Directors of NEI Parsons Ltd. for permission to publish this paper.

APPENDIX 1

Rotor stability and synchronous vibration

General

The equation for free flexural vibration of a symmetrical rotor on two pinned supports may be

written in the form
$$D_E^+(\omega^2)R = 0 \quad \text{or} \quad D_O^+(\omega^2)R = 0$$

according to whether odd (bearing forces out-of-phase) or even (bearing forces in-phase) behaviour is being considered. In these expressions R is a vibration amplitude and D^+ is a determinant which is zero when $\omega^2 = \omega_k^2$, where $\omega_k (k = 1,2,3,4$ etc) is a natural frequency on pinned supports. For example, in the case of a two mass (modified Jeffcott) model of a rotor[2]

$$D_E^+ = \omega_1^2 - \omega^2$$

and
$$D_O^+ = \omega_2^2 - \omega^2$$

In the case of a four mass rotor[2]

$$\left.\begin{aligned}
D_E^+ &= \beta(1-\beta)(\omega_3^2 - \omega^2)(\omega_1^2 - \omega^2)\\
\text{and} \quad D_O^+ &= \beta(1-\beta)(\omega_4^2 - \omega^2)(\omega_2^2 - \omega^2)
\end{aligned}\right\} \quad (A1)$$

where $\beta = \beta(\omega_1, \omega_2, \omega_3, \omega_4)$ is a number, less than unity, which indicates the proportion of the total rotor mass which is adjacent to the bearings.

Free vibration

If the journals are not pinned but are restrained only by an oil film, each may vibrate vertically with a complex amplitude Y given by

$$y - \Re[Ye^{i\omega t}]$$

and horizontally according to

$$x = \Re[Xe^{i\omega t}]$$

If the corresponding complex amplitudes of the mass adjacent to the bearing are Y_1 and X_1, the amplitude of the dynamic force sustained by the oil film at a limit of stability will be

$$\left.\begin{aligned}
K_1(Y_1 - Y) &= \frac{g}{\delta}\left[c_{11}Y + c_{12}X + \frac{i\omega}{\Omega}(b_{11}Y + b_{12}X)\right]\\
\text{and} \quad & \\
K_1(X_1 - X) &= \frac{g}{\delta}\left[c_{21}Y + c_{22}X + \frac{i\omega}{\Omega}(b_{21}Y + b_{22}X)\right]
\end{aligned}\right\} \quad (A2)$$

where mg is the steady bearing load and

$$K_1 = K_1(\omega_1, \omega_2, \omega_3, \omega_4)$$

(In the case of a Jeffcott or modified Jeffcott rotor $K_1 = \omega_1^2$).

The equations of even or odd motion can be expressed in the form

$$\left.\begin{aligned}
D_E^+(\omega^2)Y_1 - G_E(\omega^2)Y &= 0\\
D_E^+(\omega^2)X_1 - G_E(\omega^2)X &= 0
\end{aligned}\right\} \quad (A3)$$

so that, eliminating Y_1 and X_1 from equation A2

$$\begin{bmatrix} (\alpha_{11} - \Gamma^2) & \alpha_{12} \\ \alpha_{21} & (\alpha_{22} - \Gamma^2) \end{bmatrix}\begin{bmatrix} Y \\ X \end{bmatrix} = \begin{bmatrix} 0 \\ 0 \end{bmatrix} \quad (A4)$$

where
$$\alpha_{11}(\omega, \Omega) = c_{11} + i\,b_{11}\frac{\omega}{\Omega}, \text{ etc}$$

and where
$$\Gamma^2(\omega^2) = \frac{K_1\delta}{gD^+}(G^+ - D^+)$$

and the distinction between even and odd and the subscripts has been dropped for the moment.

The complex determinant
$$D^*(\omega, \Omega) = (\alpha_{11} - \Gamma^2)(\alpha_{22} - \Gamma^2) - \alpha_{12}\alpha_{21}$$

must, of course, be zero at a limit of stability. When the real and imaginary parts are each equated to zero the following results are obtained:

$$\Gamma^2 = \frac{E}{2B} \quad (A5)$$

and
$$\frac{\omega^2}{\Omega_L^2} = \frac{1}{A}\left[D - \frac{CE}{B} + \frac{E^2}{4B^2}\right] \quad (A6)$$

where
$$\begin{aligned}
A &= b_{11}b_{22} - b_{12}b_{21}\\
2B &= b_{11} + b_{22}\\
2C &= c_{11} + c_{22}\\
D &= c_{11}c_{22} - c_{12}c_{21}\\
\text{and} \quad E &= c_{11}b_{22} + c_{22}b_{11} - c_{12}b_{21} - c_{21}b_{12}
\end{aligned}$$

The eight coefficients and, therefore, the five coefficients A–E above, all vary slowly with journal speed Ω so that if they are evaluated at the rated speed and the frequency ω is obtained from equations A2 and A3, the speed Ω_L given by equation A4 will usually differ from and will, preferably, exceed the rated speed, by a significant margin.

Equation A6 shows that the lowest (even) value of the frequency ω is the most important from the point of view of flexural stability and a one mass (Jeffcott) model will be adequate for this purpose in which case
$$G^+ = \omega_1^2 \quad \text{and} \quad D^+ = \omega_1^2 - \omega^2$$

Synchronous vibration due to unbalance

Without loss of generality the response to even and odd unbalance can be calculated by considering the case of the centre of gravity of the masses adjacent to a bearing to be displaced from the axis of rotation by a distance equal to the nominal radial clearance of the bearing in which case the equations of motion can be written in the form

$$\left.\begin{aligned}
D^+Y_1 - G^+Y &= \omega^2\\
\text{and} \quad D^+X_1 - G^+X &= -i\omega^2
\end{aligned}\right\} \quad (A7)$$

where the vibration frequency ω is now equal to the angular speed Ω.

With this condition equations A2 are still applicable but equations A4 become

$$\begin{bmatrix} (\bar{\alpha}_{11} - \Gamma^2) & \bar{\alpha}_{12} \\ \bar{\alpha}_{21} & (\bar{\alpha}_{22} - \Gamma^2) \end{bmatrix} \begin{bmatrix} Y \\ X \end{bmatrix} = \begin{bmatrix} S \\ -iS \end{bmatrix} \qquad (A\,8)$$

where $\qquad \bar{\alpha}_{11} = c_{11} + i\,b_{11}$ etc

and $\qquad\qquad S = \dfrac{K_1 \delta \omega^2}{gD^+}$

The solution of equation A8 is

$$X = -(\bar{\alpha}_{21} + i\,(\bar{\alpha}_{11} - \Gamma^2))\frac{S}{\bar{D}^*}$$

and $\qquad\qquad Y = -((\bar{\alpha}_{22} - \Gamma^2) + i\bar{\alpha}_{12})\dfrac{S}{\bar{D}^*}$

where $\qquad \bar{D}^* = (\bar{\alpha}_{11} - \Gamma^2)(\bar{\alpha}_{22} - \Gamma^2) - \bar{\alpha}_{12}\bar{\alpha}_{21}$

The responses to even and odd unbalance are important and this time the frequency ω is the known, synchronous speed. Even and odd values of D^+ (and, therefore, S) and of G^+ (and, therefore, Γ^2) must be considered. The maximum amplitudes U_E and U_O can be calculated from the complex horizontal and vertical amplitudes X and Y as follows:

The displacement from equilibrium of a journal centre is given by

$$y = \Re\{Ye^{i\omega t}\}$$

and $\qquad\qquad x = \Re\{Xe^{i\omega t}\}$

ie

$$y = Y_r \cos \omega t - Y_i \sin \omega t$$
$$x = X_r \cos \omega t - X_i \sin \omega t$$

and

The time at which $\quad U = (y^2 + x^2)^{1/2} \quad$ is a maximum (or minimum) is given by

$$\tan 2\omega t_m = \frac{2(X_r X_i + Y_r Y_i)}{(X_i^2 + Y_i^2 - X_r^2 - Y_r^2)}$$

It is a simple matter to select the larger of two values of U calculated as indicated above using ωt_m and using $\omega t_m + \pi/2$.

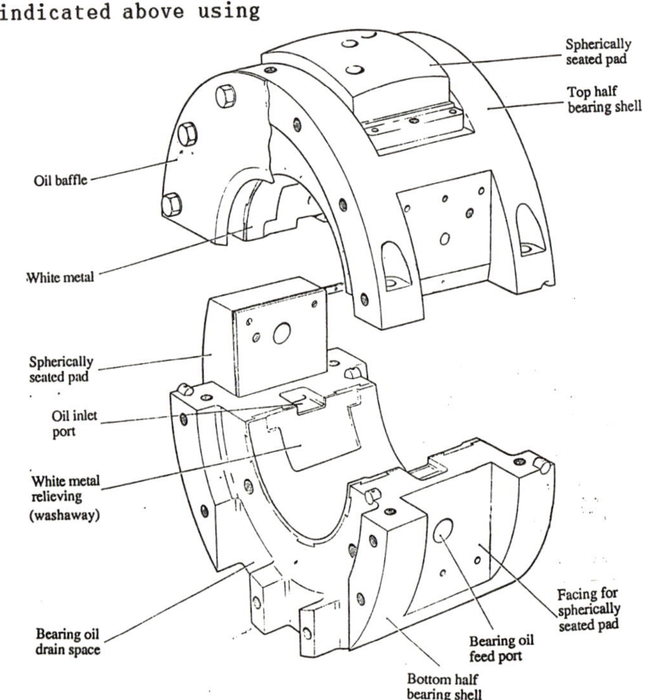

Fig 1   Elliptical journal bearing

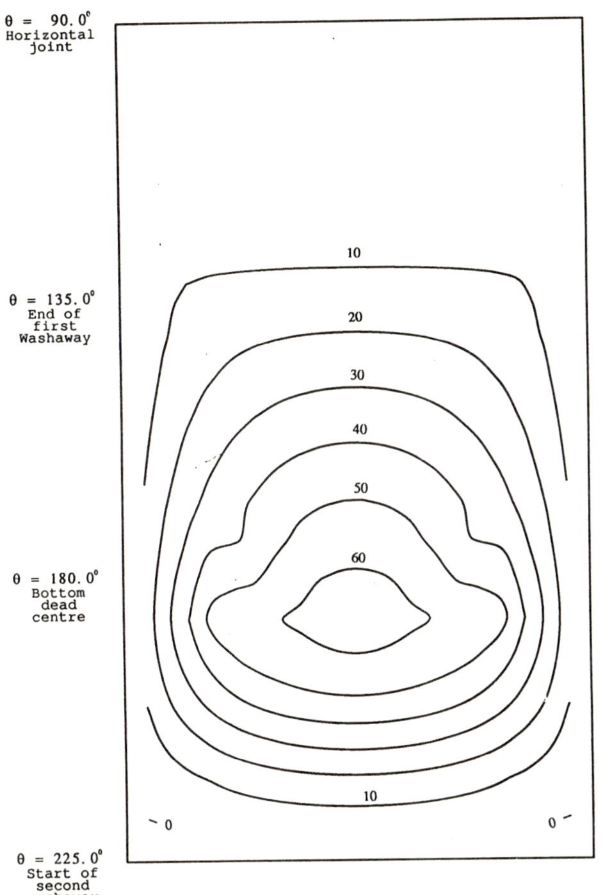

Fig 2   Pressure distribution (bar) in reference bearing

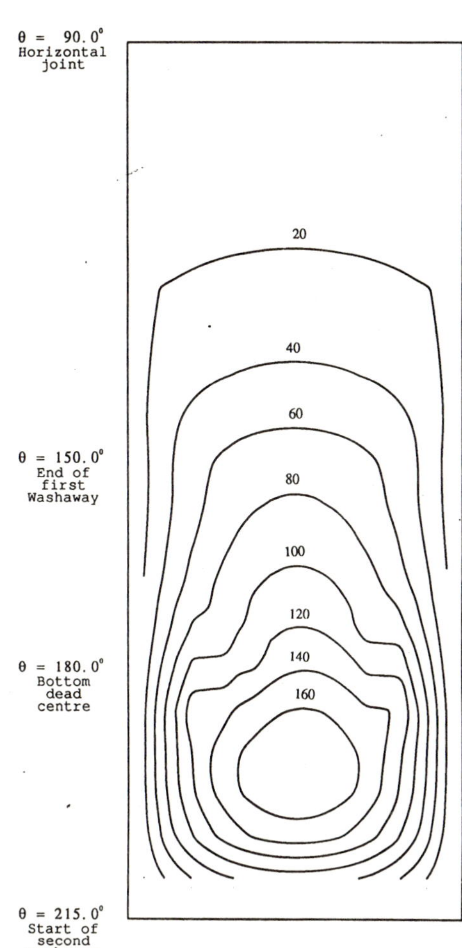

Fig 4   Pressure distribution (bar) in bearing optimized for low fuel cost

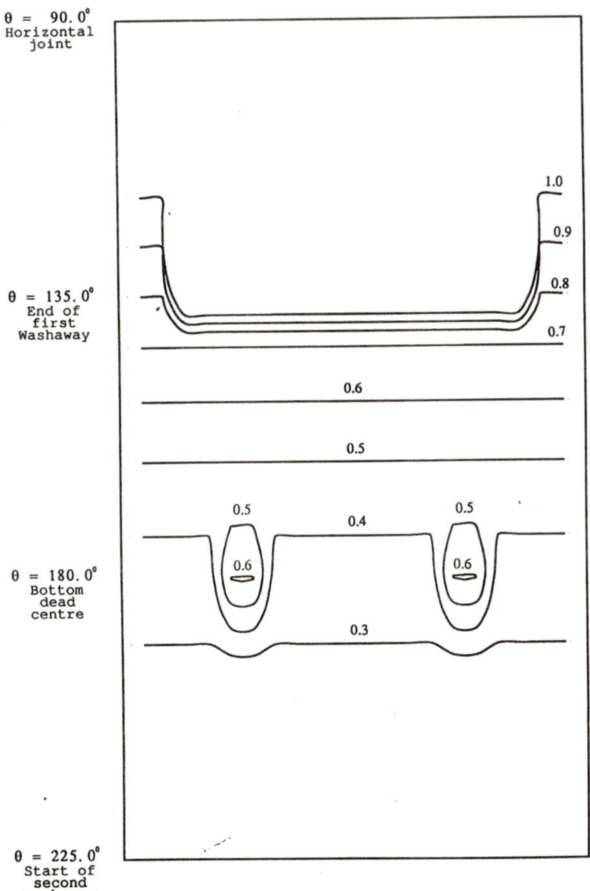

Fig 3   Film thickness (dimensionless) in reference bearing

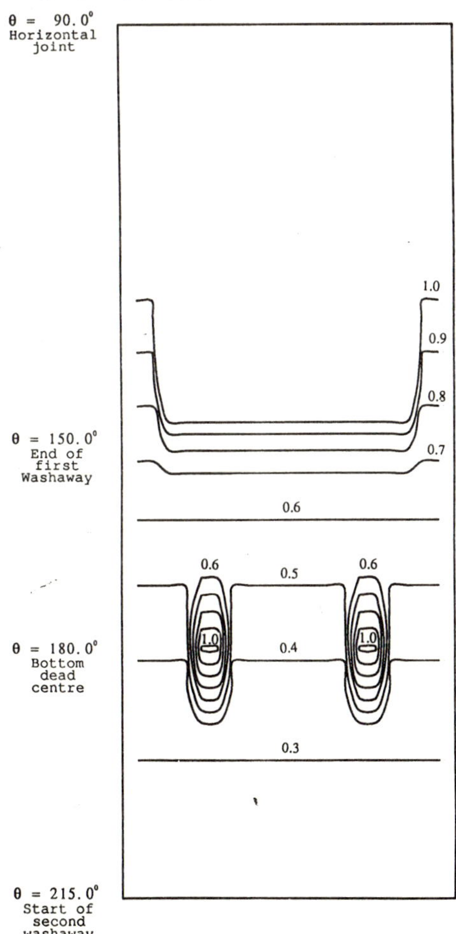

Fig 5   Film thickness (dimensionless) in bearing optimized for low fuel cost

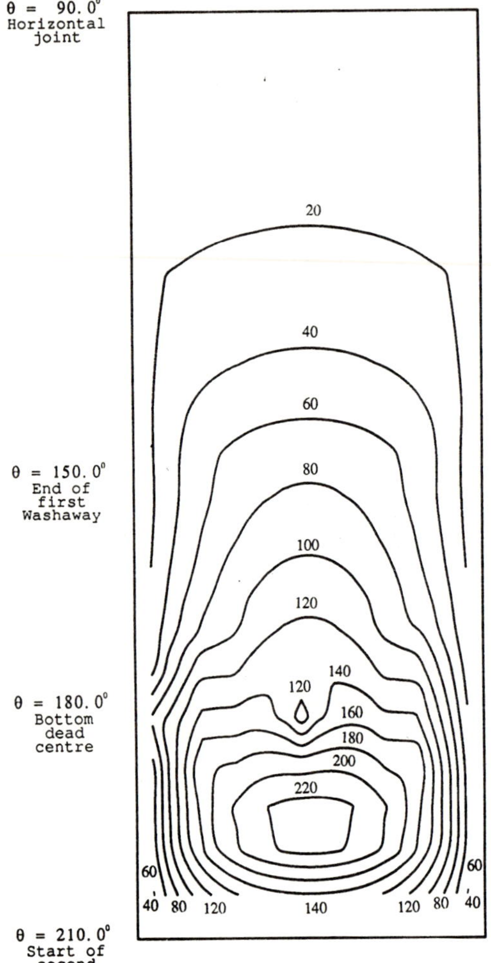

Fig 6   Pressure distribution (bar) in bearing optimized for high fuel cost

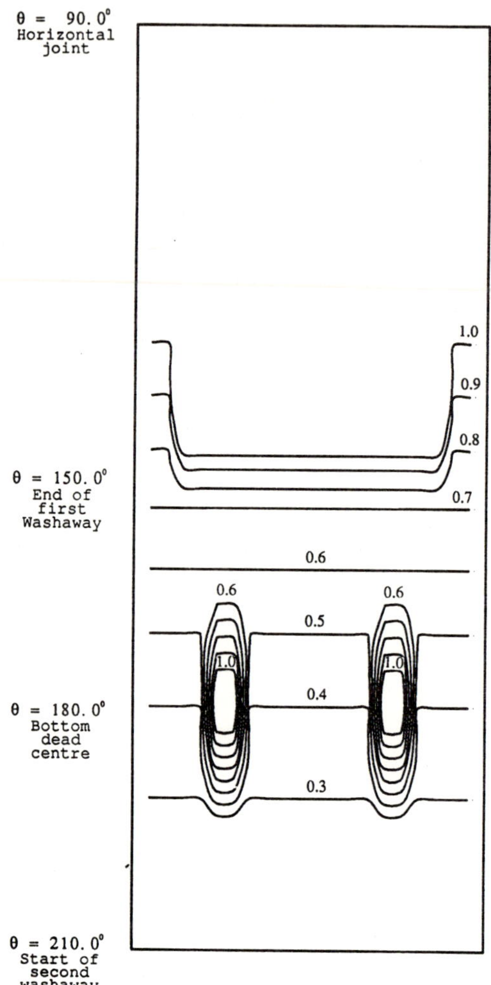

Fig 7   Film thickness (dimensionless) in bearing optimized for high fuel cost

Energy cost reduction in tilting pad thrust bearings

J E L SIMMONS, BSc, PhD, EEng, FIMechE, MIEE
Department of Mechanical Engineering, Heriot-Watt University, Edinburgh, UK
S D ADVANI, MSc, DIC, BEng
Michell Bearings, Newcastle upon Tyne, UK

SYNOPSIS Hydrodynamic, tilting pad thrust bearings are used in many industrial applications. In every case energy is absorbed by the bearing and dissipated in the form of heat. The purpose of this paper is to explore the energy consumption of thrust bearings for high speed applications under different operating conditions. It is demonstrated that a reduction in the rate at which oil is supplied can lead to significant lowering of the energy loss incurred by the bearing at no risk to its safety. Comparative experiments are reported involving offset pivot thrust pads, subject to forward and reverse rotation of the thrust collar and centre pivot thrust pads. It is found that these changes in thrust pad pivot position do not affect the energy consumed by the bearing.

1. INTRODUCTION

Oil lubricated hydrodynamic thrust bearings rely on a plentiful supply of lubricant being drawn into a convergent space thus generating a load carrying film. In many cases the supply of lubricant is guaranteed by arranging for the working faces of the bearing to be immersed in oil. This arrangement, often referred to as "flooded" lubrication, whilst very satisfactory for lower speeds, is much less suitable for high speed use since it leads to prohibitively large amounts of energy being absorbed by the bearings. Energy consumption derives from two sources; necessary frictional losses caused by shearing in the lubricating film and parasitic losses due to churning of the rim of the thrust collar in the surrounding oil. The effect of churning is not significant at low speeds, but at higher speeds typically above 40 m/s at the mean pitch diameter of the bearing, the associated energy losses increase rapidly to equal or even exceed frictional losses.

Two fundamental approaches have been devised to overcome the problem of parasitic losses and enable high speed bearings to operate successfully with acceptable power loss figures. The first method, referred to in this paper as 'shrouded' lubrication, requires that the thrust collar rim be surrounded by a shroud or baffles. The presence of a restriction at the collar rim substantially reduces the amount of possible churning. The second basic approach, known as 'low loss' or 'directed' lubrication, is a very simple one in which the rim of the collar runs free of oil and churning losses are virtually eliminated.

There have been a number of published reports in which the two design philosophies are described (1,2,3,4,5). Energy consumption in a bearing is manifested principally by the temperature rise between supply and drain of the oil which passes through the system. The ability of energy efficient design to limit this increase when compared with conventional, flooded lubrication has been well established by previous investigations.

The main conclusion of a recently published report (6) on a direct comparison between the two design philosophies was that the low loss approach was the more energy efficient by about 25 per cent when operating conditions were the same for both configurations. Because the principle of the low loss bearing is a very simple one, it was suggested in the same report that bearings of this sort are the most cost-effective arrangement currently available for high speed, energy efficient thrust bearings.

This paper reports the results of some further experiments with an example low loss bearing which have been carried out to determine the beneficial effects in energy terms of changing its operating conditions. In particular the bearing was operated with substantially reduced oil flow and with offset pivot and with centre pivot thrust pads.

2. EXPERIMENTAL APPARATUS

The experiments were carried out using a double thrust bearing selected from the middle of a standard range. Each thrust face comprised eight tilting thrust pads supported by a ring of load equalising segments of conventional design. Basic dimensional information about the bearing is given by Table 1. The bearing was assembled in a new casing mounted on an existing, high speed, horizontal bearing rig which has been described on a previous occasion (1). The bearing configuration and lubrication arrangements are shown by Figure 1.

Oil is supplied under positive pressure from an external source to an annulus around the circumference of the retaining ring which contains the equalising segments. Thence it is directed on to the thrust collar through orifices inserted in the retaining ring. In the low loss bearing the diameter and number of these orifices are used to meter the flow of oil to the thrust surfaces. From a design point of view it is easy to cater for any duty simply by changing the size and/or the number of holes. Overall size of bearing represents no limitation since the number of holes can be increased to ensure adequate supply across the full width of the thrust pad. In the bearing used in the experiments described in this paper there were two orifices, each 3 m.m. in diameter, between each pair of thrust pads. The emergent jets of oil are directed on to the surface of the thrust collar and a proportion of lubricant is drawn into the hydrodynamic films between the thrust pads and collar. Lubricant is discharged from the bearing simply by being allowed to fall, unrestricted, to drain by way of the large opening in the lower half casing.

3. OPERATING CONDITIONS

Axial load was applied to the test bearing by a series of hydraulic pistons positioned behind the retaining ring of a completely separate loading bearing adjacent to the test module. The effect of pressurising the hydraulic cylinders in the loading bearing was to create an axial force on the shaft which was absorbed by one of the faces of the test bearing. The main series of experiments reported in this paper, were performed at specific loads of 2 MPa and 4 MPa.

The d.c. motor and thyristor drive of the experimental rig meant that speed was continuously variable with maximum speed being limited only by the power available. However, since the energy savings mechanisms are most effective at higher speeds experiments were conducted at three speeds, 5000 rev/min, 7500 rev/min and 10,000 rev/min, typical of high speed equipment such as compressors, turbines and gearboxes. These rotational speeds are equivalent to sliding velocities at the mean pitch diameter of the thrust ring of 47.3 m/s, 70.9 m/s and 94.6 m/s respectively.

The lubricant used throughout the experiments was a standard ISO VG 32 turbine oil supplied to the casing at a nominal 45°C. In practice experiments took place with oil supply temperatures between 44°C and 46.5°C. It is usual to base recommended rates for oil supply on the expected rise in temperature between the supply point and the drain point from the bearing casing. Values of lubricant temperature increase up to about 17°C are common industrial practice. In the current experiments the oil flow rate was reduced at each operating speed and load from an initial amount to a rate about half that originally supplied with a consequent increase in oil drain temperature. The reductions in oil flow were carried out in stages and the experimental results are reported below.

The oil flow reduction experiments were carried out when the bearing being examined was fitted with offset pivot pads. Pivot position was 0.6 of the distance between the leading and trailing edges of each pad as shown by Figure 2. Subsequently the bearing was operated in reverse for extended periods at the 2 MPa and 4 MPa loadings, with a reduced oil supply. Following this the offset pivot pads were replaced by centre pivot pads and the experiment repeated.

4. INSTRUMENTATION

The temperatures of oil supplied to and leaving the bearing were monitored by thermocouples at inlet and drain respectively. Oil flow rate was measured using a calibrated, variable orifice meter placed in the supply line. These values enabled power absorbed by the test bearing to be calculated by multiplying flow rate and temperature difference (T_{diff}) between supply and outlet with the specific heat of the oil concerned.

The most important and widely used parameter in monitoring hydrodynamic thrust bearing safety is maximum thrust pad temperature. This temperature (T_{max}) was measured by thermocouples embedded 3 m.m. below the whitemetal (babbitt) surface at the centre of the outboard trailing quadrant formed between the mean pitch diameter and the pad pivot as shown by Figure 2 for the offset pivot pad. This location is in agreement with the suggestions of other workers in this field (7,8) and its suitability has been discussed in an earlier paper (1) specifically concerned with measurements of maximum temperature in tilting pad thrust bearings. All the loaded pads were instrumented in the same way and the values given in the results are the arithmetic mean of the readings taken from the eight pads on the loaded side of the bearing.

In the experimental work thermocouple temperatures were recorded continuously by pen plotter. The time it took for values to settle after each change of duty was about 15 minutes. Thereafter the given conditions were maintained for about 1½ hours to ensure that the readings were properly stabilised and not subject to long term variations.

5. EXPERIMENTAL RESULTS

5.1 <u>Reduction of oil flow with offset pivot pads</u>

Figures 3 and 4 show maximum pad temperature, T_{max}, and temperature difference between supply and drain, T_{diff}, plotted against oil flow for the test bearing at the 2 MPa and 4 MPa specific loadings respectively. In each case as the initial oil flow is reduced by up to about half there is a gradual increase in both maximum pad temperature and in the temperature rise of the oil between supply and drain. Maximum pad temperature remains well within acceptable and safe limits for whitemetal thrust pad surface material. The temperature of the lubricant as it passes through the bearing increases from initial values lying between 10°C and 18°C to a maximum value, recorded for the 4 MPa load at 10000 rev/min of 28°C. The bearing performed well throughout these experiments. When it was stripped down subsequently there was no sign of any distress to any of the components and the thrust pad surface material appeared in good condition.

The 2 MPa results of Figure 3 are repeated as broken lines on Figure 4 so that the effect of increasing load on T_{max} and T_{diff} can be seen. For a given speed and oil flow rate T_{max} increased by 20°C to 25°C for the specific load change from 2 MPa to 4 MPa. T_{diff} increases by a very much smaller amount in absolute terms but one which is still significant as a proportion of the 2 MPa value. The change in T_{diff} with load is reflected in Figure 5 which shows power absorbed by the low loss bearing against oil supply rate for both specific loadings. Results are plotted for the three speeds at which experiments were carried out. At each speed approximately 15 per cent more energy is absorbed when the load is doubled. At the lowest speed, 5000 r.p.m., there is little variation in power loss with change in oil flow. However as speed increases the change in power loss becomes more significant. At 10000 r.p.m. the decrease in power loss is approximately 15 per cent at both 2 MPa and 4 MPa loadings given a 40 per cent reduction in oil flow.

5.2   Reverse rotation with offset pivot pads

The major part of the experimental work described in this paper was carried out using bearings with thrust pads which had offset pivots. It is well-known that offset pivot pads run significantly cooler (1) than equivalent centre pivot pads given the same operating conditions. Because of this advantage offset pivot pads are a frequent choice for machinery in which the duty requirement calls for a single direction of rotation. However sometimes such machinery may be called upon to act in reverse. Example situations may be at machine rundown under pumped load or under certain emergency conditions. In these circumstances the reverse running capacity of offset pivot pads is important.

In order to establish the reverse running capability of the experimental bearing, it was operated in reverse for the standard 2 MPa and 4 MPa loads over the full range of speeds using one set of oil flows reduced by between 15 per cent and 30 per cent compared with the starting point of the experiments described above. The bearing performed satisfactorily in every way. The results of the reverse rotation experiments are given by Table 2 together with the results of the equivalent series of forward rotation experiments. It can be seen that the maximum pad temperature for reverse rotation was some 20°C higher than that for forward rotation. Nevertheless, the maximum pad temperature remained within acceptable bounds for such an extreme situation. The temperature difference between supply and drain did not increase with reverse rotation and consequently the power absorbed by the bearing shows no significant difference when operated in reverse.

5.3   Comparison with centre pivot pads

The advantage of offset pivot pads in comparison with centre point pads is borne out by the final set of results given in Table 2. The difference in the value of T_{max} for the offset pad in forward rotation and the centre pivot pad is about 10° for the 2 MPa loading and 20°C for the higher 4 MPa load. These results compare well with those quoted in reference (1) for the difference to be expected between offset and centre pivoted pads.

Examination of the values of T_{diff} for the bearing when fitted with centre pivot pads shows that there is almost no difference between the values recorded and those recorded for the bearing when the offset pads were fitted. Hence the energy absorbed is the same notwithstanding the much higher maximum pad temperature in the case of the centre pivot pad bearing.

It is interesting to compare the centre pivot pad results with those of the offset pad in reverse rotation. In each case as already indicated there is little or no difference in the values of T_{diff}. There is a difference however in maximum pad temperature. At the lower, 2 MPa applied load the centre pivot pad is 6°C to 12°C cooler than the offset pad operating in reverse. However when the 4 MPa load is applied, there is remarkable similarity between the results.

6.   DISCUSSION

6.1   Effect of oil flow

Figure 5 shows the effect of decreasing oil supply rate on energy consumed by the bearing. In every case the reduction in oil supply leads to decrease in power absorbed. At 5000 rev/min there is little change when oil supply is reduced while at 10000 rev/min there is a significant decrease in energy loss given a substantial reduction in oil supply.

These results of the experiments with reduced oil flow demonstrate the tolerance of low loss bearings to a decrease in oil supply. Maximum pad temperatures certainly increased gradually with diminishing oil flow rate but not to unsafe levels. Temperature difference between supply and drain also increased but again not to a hazardous extent. Considerable experience with other similar bearings suggests that there should be no danger in operating for long periods with temperature differences up to 30°C provided oil is supplied at about 45°C and maximum pad temperatures do not exceed 130°C. Figure 5 shows that as oil supply was reduced power loss decreased. It is proposed that this is an indication that hydrodynamic lubrication persisted throughout the experiments. If this were not the case then there is likely to have been an increase in power loss to accompany the break up of the fluid film. The actual minimum oil supply level when this would have occurred as a precursor to failure exists at some point beyond the range of the present work. The nature of fluid film bearings makes the experimental or theoretical determination of final limits of this sort very difficult. What the experimental work has done, however, has been to increase knowledge of the operation of low loss bearings beyond what are present day normal boundaries for industrial applications.

6.2   Effect of pivot position

The results given in Table 2 show that for the experimental bearing and the operating conditions chosen, there is little significant variation in temperature difference between supply and drain resulting from change in thrust pad pivot position. In other words the energy absorbed by the bearing is the same whether offset pivot pads are used, in forward or reverse rotation, or if centre pivot pads are used.

The offset pivot pads, in forward rotation, offer values of maximum pad temperature which are substantially less than those of the centre pivot pads. The experiments carried out with the offset pivot pads in reverse show that they operate satisfactorily to provide acceptable values of T_{max} and T_{diff}. At the 4 MPa loading, the results for the offset pivot pads in reverse are no worse than for the centre pivot pads. This suggests that for all applications in which there is a predominant direction of rotation there is every reason to select offset pivot pads and take advantage of the reduced T_{max} in comparison with centre pivot pads. Centre pivot pad bearings remain an appropriate design choice for applications in which there is an equal requirement for rotation in both directions.

It was interesting to note the effect of applied load on the maximum temperatures of offset pivot pads operating in reverse in comparison with centre pivot pads.

We summarise that the performance of offset pads in reverse is aided by their ability to bend under the effects of load and temperature to form the necessary convergent hydrodynamic film in the changed direction of rotation. Evidently pad bending is related to the size of the applied load. In the case of the 4 MPA loading it seems that the deflection is such that the offset pad operating in reverse acts in a very similar way to a conventional centre pivot pad bearing.

6.3 Comparison with theory and previous experimental work

The minimum, unavoidable power loss in a hydrodynamic bearing is that due to shearing the fluid flims between pads and collar. In the case of a double thrust bearing the majority of losses are incurred on a loaded face but nevertheless a small and significant proportion is due also to the unloaded face.

Theoretical results were obtained for both faces for conditions corresponding to the current experiments using programs developed at Michell Bearings based on established design principles. The theoretical results for the 4 MPa load are shown on Figure 6. Comparison between theory and experiment shows good agreement and supports the idea that this design is one in which parasitic losses have been effectively eliminated. The considerable size of the parasitic losses in comparison with the fluid film losses can be seen from the theoretical energy consumption of a fully flooded bearing which is also shown on Figure 6. The flooded bearing includes both shear losses from the fluid film and parasitic losses.

A previous paper (6) included results for a "shrouded" energy saving bearing in direct comparison with the low loss bearing described in this paper. The results for the shrouded bearing, given on Figure 6, show how this design approach represents a significant improvement on the energy cost of a fully flooded bearing. Nevertheless on energy cost alone the shrouded bearing remains approximately 25 per cent more expensive than the low loss bearing for the same operating conditions.

7. CONCLUSIONS

The energy absorbed by the low-loss bearing conforms closely with the theoretical minimum loss expected due to shearing of the hydrodynamic films on the loaded and unloaed faces. This is considerably less than the loss which would be incurred by a fully flooded bearing and, as shown by previous experimental work, significantly less than that incurred by an equivalent shrouded bearing.

The low loss bearing has been shown to be extremely tolerant of major reductions in oil supply. It has been found that reducing oil supply by about 40 per cent resulted in the energy absorbed by the bearing reducing by up to 15 per cent. The greatest reduction occurred at the highest speeds of operation.

The ability of offset pivot pad bearings to run satisfactorily in reverse has been demonstrated and a comparison with an equivalent centre pivot pad bearing has been carried out. It was found that the energy absorbed by the low loss bearing is independent of thrust pad pivot position.

In conclusion it is suggested that the offset pivot pad, low loss bearing operating with a reduced lubricant supply represents the optimum energy efficient choice currently available to designers for all high speed operations in which there is a predominant direction of rotation.

8. ACKNOWLEDGEMENTS

The authors are grateful to the directors of Vickers Plc - Michell Bearings for permission to publish this paper and to many members of staff of the company for their assistance. Particular thanks are due to Mr J Wilkin for his contribution to the experimental work reported.

REFERENCES

(1) Horner, D., Simmons J.E.L., and Advani, S.D. 'Measurements of Maximum Temperaturesin Tilting-Pad Thrust Bearings.' Tribology Transactions, Jan 1988, vol. 31, pp 44-53.

(2) Gregory, R.S. 'Factors influencing power loss of tilting-pad bearings.' Trans ASME, Journal of Lubrication Technology, April 1979, vol 101, pp 154-163.

(3) New, N.H. 'Experimental Comparison of flooded, directed and inlet orifice type of lubrication for a tilting pad bearing.' Trans ASME, Journal of Lubrication Technology, January 1974, pp 22-27.

(4) New, N.H. 'Comparison of flooded and directed lubrication tilting pad thrust bearings.' Tribology International, 1979, vol 12, pp 251-254.

(5) Mikula, A.M. and Gregory, R.S. 'A comparison of tilting pad thrust bearing lubricant supply methods'. Trans ASME, Journal of Lubrication Technology, Jan 1983, vol. 105, pp 39-47.

(6) Simmons, J.E.L. and Advani, S.D. 'The comparative performance of energy efficient, tilting pad thrust bearings for high speed appliction'. Proc. Annual Meeting STLE, Montreal, Canada. April 29-May 2, 1991.

(7) Capitao, J.W., Gregory, R.S. and Whitford, R.P. 'Effects of high-operating speeds on tilting thrust pad performance'. Trans ASME Journal of Lubrication Technology, 1976, vol 98, pp 73-80.

(8) Mikula, A.M. 'Evaluating Tilting Pad Thrust Bearing Operating Temperatures'. Proc. ASLE Annual Meeting, Las Vegas, Nevada. May 6-9, 1985. Paper 85-AM-IE-1.

(9) Mikula, A.M. 'The leading-edge-groove tilting-pad thrust bearing.' Trans ASME, Journal of Tribology, July 1985, vol 107, pp 423-430.

Table 1 Dimensional information for the low loss experimental bearing

Number of Pads per Thrust Face	8
Inside Diameter	114.3 mm
Outside Diameter	228.6 mm
Surface Area per Thrust Face	25994 mm^2
Thrust Collar Diameter	231.7 mm
Shaft Diameter	100 mm
Axial Clearance in Bearings	0.7 mm
Area of Drain	7650 mm^2
Number of housing inlets per thrust face	2
Diameter	20 mm
Number of orifices per pad	2
Diameter	3 mm

Table 2 Operating conditions and results for the test bearing under forward and reverse rotation when fitted with offset pivot pads, and when fitted with centre pivot pads

			Offset Pivots				Centre Pivots	
Speed	Specific Load	Total Oil Flow	Forward		Reverse			
			T_{max}	T_{diff}	T_{max}	T_{diff}	T_{max}	T_{diff}
rev/min	MPa	l/hr	°C	°C	°C	°C	°C	°C
5000	2.0	1920	90	13.5	106	14	100	14.5
7500	2.0	2840	95	18.5	117	19	107	19.5
10000	2.0	3960	100	21	122	23.5	110	24
5000	4.0	2240	106	13.5	113	14	116	15.5
7500	4.0	3300	116	19	128	18	130	19
10000	4.0	4520	115	21	138	21.5	137	22

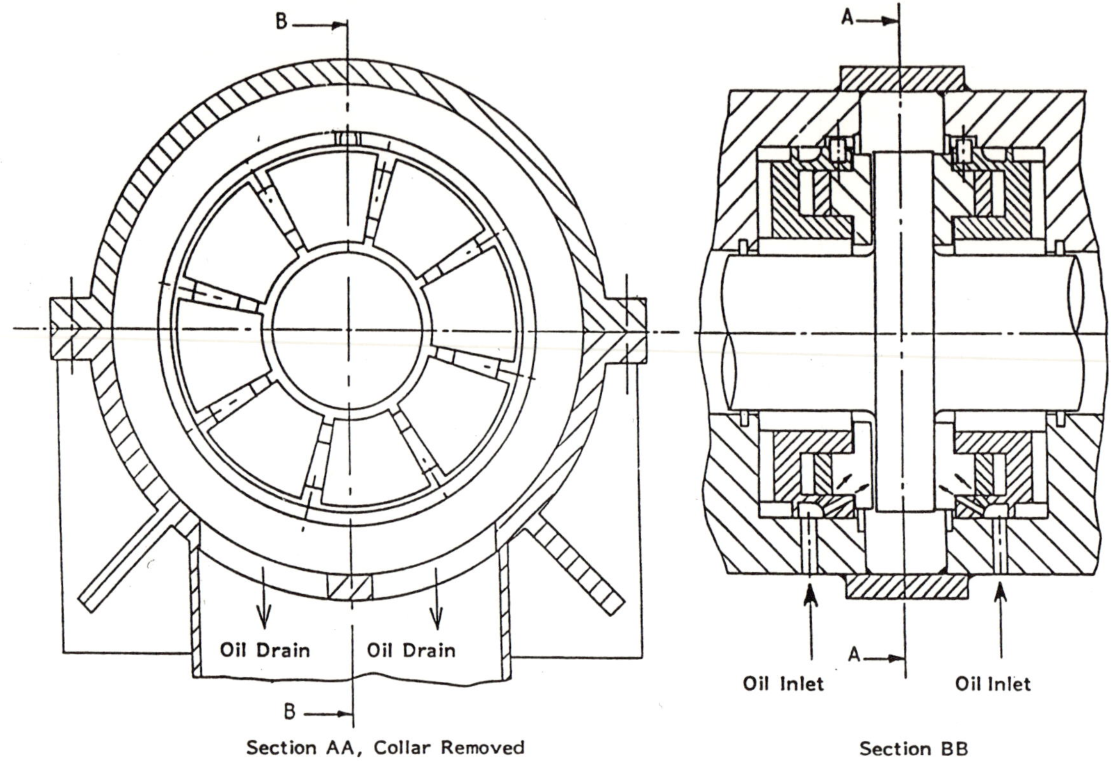

Fig 1   Experimental bearing

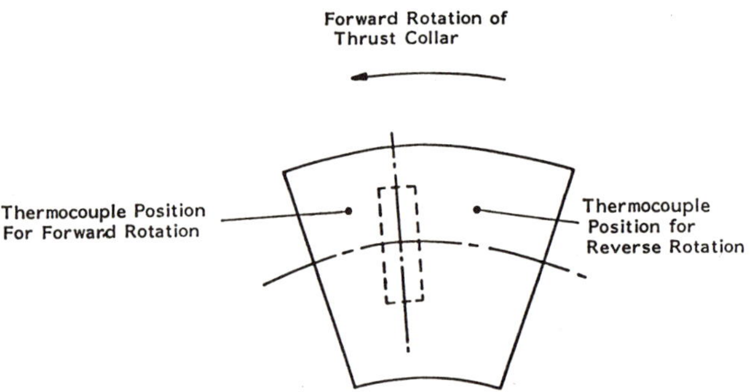

Fig 2   Sketch showing location of thermocouples measuring Tmax for forward and reverse rotation

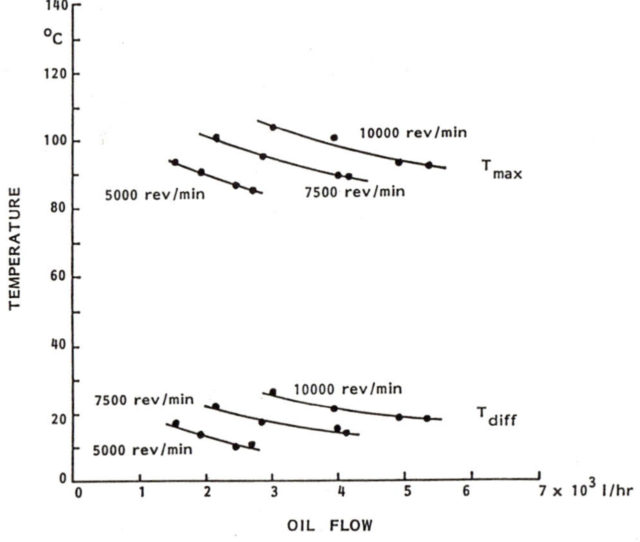

Fig 3   Variation of maximum pad temperature and temperature difference with oil flow; 2MPa load

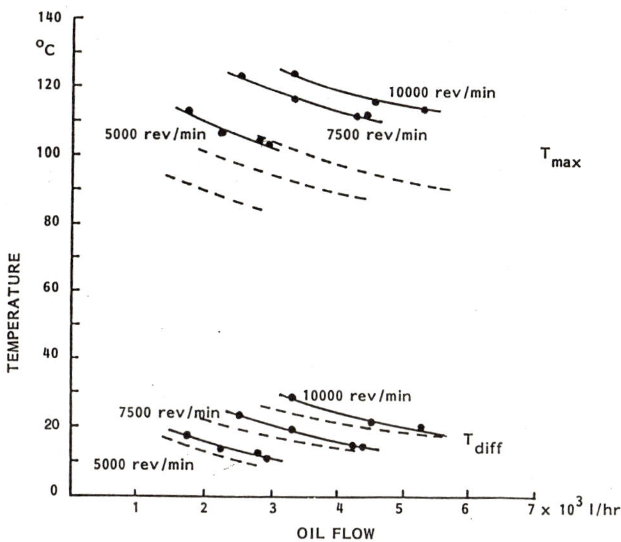

Fig 4   Variation of maximum pad temperature and temperature difference with oil flow; 4MPa load (--- 2 MPa results from fig.3)

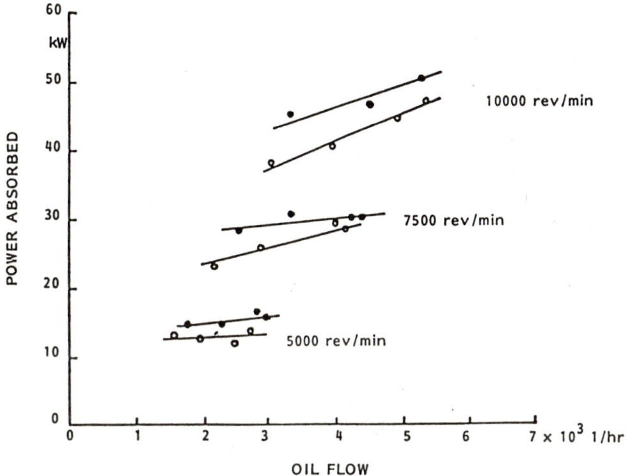

Fig 5   Variation of power absorbed by bearing with oil flow; o, 2MPa load; ●, 4MPa load

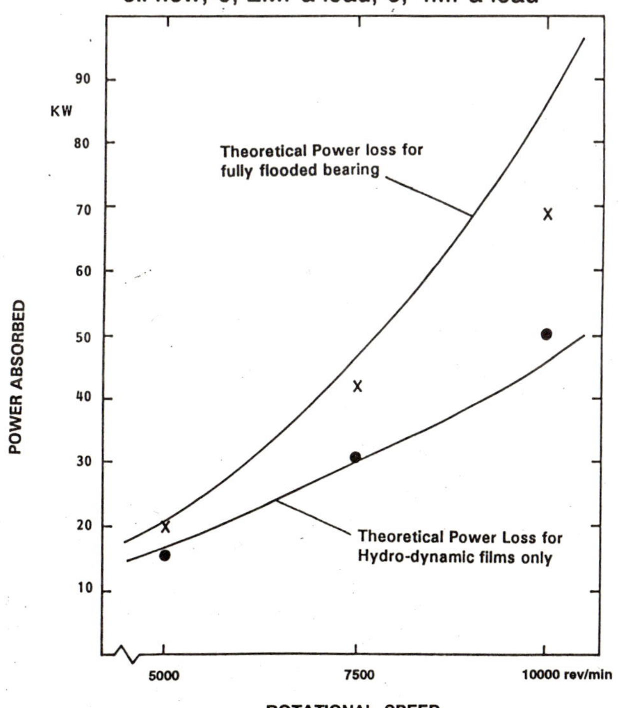

Fig 6   Variation of power absorbed with speed; comparison of experimental and theroretical results; 4MPa load (Experimental results; ●, low loss bearing; x, shrouded bearing, ref.6)

Water lubricated ceramic thrust bearing-design study

G S RITCHIE, BSc
Engineering Research Centre, Whetstone, UK

SYNOPSIS   Using ESDU item 83004, "Calculation methods for steadily loaded offset pivot tilting-pad thrust bearings", the design implications of a water lubricated thrust bearing under high specific loading (as suggested in the Glacier-Kemel bearing) are examined.

While accepting that film thicknesses will be very small, it is predicted that power loss and maximum temperatures are also extremely low compared with those of a conventional oil-lubricated bearing. The dominant role of conduction as the heat removal process from the lubricating film in such bearings is examined.

Some experimental data supporting the findings is presented.

1   INTRODUCTION

ESDU item 83004 (ref. 1) is a guide giving a calculation scheme enabling a designer to arrive at an appropriate design of offset pivoted tilting-pad thrust bearing. The analysis is, to a great extent, based on theoretical and experimental work by Neal (ref. 2), and, indeed, example 1 of ref. 1 relates to a bearing closely resembling one of the test bearings of Neal's experiments.

The major purpose of this presentation is to explore the design process of ESDU 83004 for a water-lubricated bearing with ceramic bearing components, in the light of the claimed large load capacity of such bearings by Kamelmacher (ref. 3), using silicon carbide. For conventional oil-lubricated white-metal bearings ref. 1 recommends a maximum specific loading of 2MPa; on the other hand ref. 3 claims allowable specific loading approaching ten times this value in combination with low viscosity lubricants such as water. Thus film thicknesses certainly an order of magnitude smaller than those considered 'acceptable' in conventional bearings must be anticipated.

To set the scene the ESDU 83004 process is applied first to a conventional oil lubricated bearing with a design specific loading of 2MPa, and subsequently to a water lubricated bearing with a design specific loading of 15MPa.

2   OIL LUBRICATED BEARING

Example 1 of ref. 1 calls for the design of an offset tilting pad bearing to carry a load of 22kN at a speed of 6000 rpm. The lubricant is constrained to be an ISO VG22 mineral oil with a supply temperature of 50°C. The shaft diameter is 0.076m. The inner diameter d is assumed to be 0.089m, and the design specific load is 2MPa.

The design procedure gives:

Required bearing area =
$22000/2.0_{10}6 = 0.011\text{m}^2$

Outer diameter D for a full pad complement is calculated from:

$$0.75 \times \frac{\pi}{4} (D^2 - 0.089^2) = 0.011$$

giving D = 0.163 m and D/d = 1.83

Mean diameter	dm	= 0.126m
Pad breadth	b	= 0.037m
Pad length	L	= 0.037m
Number of pads	n	= 8

Actual specific load =
$W/nbL = 2.01_{10}6 \text{ N/m}^2$

In order to calculate the bearing parameters it is necessary, in general, to perform an iteration on the mean effective viscosity.

For this exercise, the iterated values will be used, which gives:

Effective temperature	T_e	= 66°C
Effective viscosity	η_{e}	= 0.008 Ns/m^2

The operating minimum film thickness is calculated from:

$$h = 0.46 \left\{ \frac{\eta_e N\, d_m b}{P_{mean}} \right\}^{\frac{1}{2}} \qquad (N \text{ in rps})$$

$= 19.8_{10}\text{-6m}$, an adequate value.

The bearing temperature conditions involve a film Péclet number in the form:

$$P_e = \frac{8.9\ Nd_m\, h^2}{\kappa\, L}$$

where κ is the thermal diffusivity of the oil, here taken to be $0.08_{10}\text{-6 m}^2\text{/s}$.

Hence $P_e = 14.9$

The maximum and mean temperature rises above the feed temperature T_f (=50°C) are calculated via:

$$T_{max} - T_f = \frac{23}{\varrho C(1+S_1)} \cdot \left\{ \frac{P_e}{P_e + \alpha} \right\} \frac{L}{b} P_{mean}$$

and

$$T_e - T_f = \frac{T_{max} - T_f}{S_2}$$

S_1 is a factor relating to relative proportions of the runner below a pad and that exposed to cooling oil, with, in this case the suggested value is 0.12.

S_2 is a factor relating to the speed with an empirical value in this case of 1.40.

α is a factor which, in the bracket $(P_e/(P_e + \alpha))$ gives an indication of the proportion of heat removed by convected oil, the remainder being conducted into the runner and pads. The suggested value is $\alpha = 1$.

Hence with the thermal capacity of the oil assumed to be $1.7_{10}6 J/m^3 K$

$$T_{max} - T_f = 22.8K$$

$$T_e - T_f = 16.3K \quad (i.e. \ T_e = 66°C,$$
the assumed value)

The power loss is calculated from

$$H = \frac{30hNd_mW}{b} = 4450W$$

The pad inflow, for eight pads, is given by

$$Q_i = 3.3 \ nN \ d_m \ b \ h = 244_{10}-6 \ m^3/s$$

At this minimum flow rate the mean outlet temperature T_0 is given by

$$T_0 = T_f + \frac{H}{\varrho C Q_i} = 60.7°C$$

This completes the design exercise of the conventional bearing, resulting in a perfectly acceptable design.

3   WATER-LUBRICATED BEARING

For this bearing the changes adopted are the use of water at 50°C in a flooded bearing and that the allowable specific load is increased to $15_{10}6$ MPa.

The outer diameter for a full complement of pads, retaining d = 0.089m, is calculated from:

$$0.75 \times \frac{\pi}{4} (D^2 - 0.089^2) = \frac{22000}{15 \cdot 10^6}$$

Whence D = 0.102m, b = L = 0.0065m

$$D/d = 1.146 \qquad n > 30$$

Since it is clearly highly undesirable and impractical to have such a large number of such small pads, resort is made to a reduced

complement bearing. If an eight pad arrangement is chosen, which, with the allowable specific load, indicates that b = L = 0.0135m, so that

$$D = 0.116m; \ dm = 0.1025m$$

Assuming that $T_f = 53°C$ and that $T_e - T_f = 2K$, $T_e = 55°C$ and $\eta_e = 0.0005 Ns/m^2$, the minimum film thickness is calculated to be

$$h = 0.99_{10}-6m$$

Accepting that such a low value is inevitable, the temperature calculation requires a film Péclet number which, with a thermal diffusivity for water of $0.16_{10}-6m^2/s$, yields

$$P_e = 0.041$$

For the temperature calculations, since there is only a 45% complement of pads, $S_1 = 0.40$, and the thermal capacity of water is $\varrho C = 4.1_{10}6 J/m^3K$.

Hence $T_{max} - T_f = 2.4K$

$$T_e - T_f = 1.7K \quad (i.e. \ T_e = 55°C$$
the assumed value)

The power loss is calculated to be H = 496W; the bearing inflow $3.7_{10}-6m^3/s$ and the outlet temperature based on this minimum flow 84°C. Such a temperature is incompatible with the assumed film temperatures so that a very significantly higher flow must be maintained through the bearing housing to maintain $T_f = 53°C$. The increased flow may be calculated to be $40_{10}-6m^3/s$, ie. still a low value.

The predicted film thickness of the order of $1 \mu m$ demands a bearing assembly wherein the pads must be self-adjusting to the plane of the collar within very tight limits, either by a mechanical linkage or by controlled elastic distortion. It is not absolutely clear that the ESDU requirement of component finish of Ra = h/20 remains valid but with lapping of silicon carbide this may certainly be achieved.

Hence on perfunctory examination it seems that the high specific load water-lubricated bearing may operate with extremely modest temperature rises assuming that the ESDU model is still reliable. In particular the small temperature rises are predicted as a direct result of the small Péclet number where heat transfer by conduction to the collar is, in this case, given as 96% of the total. This depends less upon the accuracy of the Péclet number which is proportional to the square of the film thickness, than upon the value assumed for α, with its suggested value of unity. The crucial question is whether the requisite quantity of heat can be conducted from the lubricant film into the runner, thence out of the runner to the cooling water with the necessary small temperature differences.

In the appendix of ref. 2, Neal considers the problem of heat flow into the runner as it traverses a pad, giving a surface temperature rise and a depth over which the temperature is assumed to drop linearly to a bulk runner temperature.

Using the notation of ref. 2:

$$\Delta T_{max} = \frac{2q}{\lambda}\left\{\frac{\kappa l}{\pi U}\right\}^{\frac{1}{2}} \; ; \quad \delta = \left\{\frac{\pi \kappa l}{U}\right\}^{\frac{1}{2}}$$

where in the case here

q - heat flux
$(= 496/(8 \times 0.0135^2) = 0.34_{10}6 \; W/m^2)$

λ - runner conductivity
(= 100 W/mK for SiC)

κ - runner thermal diffusivity
$(= 48_{10}\text{-}6 \; m^2/s \text{ for SiC})$

U - runner velocity (= 32.2 m/s)

$\Delta T_{max} = 0.56K; \quad \delta = 0.25_{10}\text{-}3 \; m$

Hence the bulk runner temperature is 0.56K below its surface temperature and the penetration depth is small.

The heat flux must be removed from the runner by the lubricant in the interpad gaps. If it is assumed that developed turbulent shear flow exists then an appropriate Nusselt number (ref. 4) is

$$Nu = 0.024 \; Re^{0.8} Pr^{0.6}$$

At the assumed condition the Reynolds number $Re \sim 3.1_{10}6$ and the Prandtl number $Pr \sim 3.7$, whence the non-dimensional heat transfer coefficient $Nu \sim 8000$. Since $Nu = qd_m/2 \Delta T$ the dimensional heat transfer coefficient htc is given by:

$$htc = 94000 \; W/m^2K$$

The cooled area may be computed as:

$$A = \frac{\pi}{4}(D^2 - d^2) - 8 \, L \, b = 2.89_{10}\text{-}3 \; m^2$$

Finally the temperature drop to the lubricant is given by:

$$T = 511/(2.89_{10}\text{-}3 \times 94000) = 1.88K$$

It may be concluded from these somewhat speculative calculations that the conduction route for heat removal to the throughflow of lubricant is very effective in the design case in producing low temperature rises of the same order as those predicted by the ESDU procedure, provided only that hydrodynamic lubrication conditions exist under the pads.

4   SOME EXPERIMENTAL RESULTS

Figure 1 shows a cross section through the test head of the seawater-lubricated thrust bearing rig at the Engineering Research Centre in which bearing loads of +30kN may be applied to a test bearing at speeds variable between 300-4000 rpm in either seawater or fresh water. The arrangement necessary to fit a Glacier-Kemel M-type 18-pad bearing, and alternatively a non-standard N-type 18-pad bearing is illustrated. The standard rig instrumentation system automatically acquires speed, shaft torque, inlet flow rate, inlet temperature, outlet temperature, vessel pressure, bearing differential pressure and net bearing load. The sole additional bearing instrumentation was sheathed thermocouples in predrilled axial holes in the silicon-carbide thrust pads at the pivot centre-line, which can be seen in figure 2. These thermocouples were retained by an appropriate adhesive which proved to be perfectly adequate in locating and sealing the thermocouples.

The overall program consisted of a very comprehensive range of tests of offset/centre pivoted bearings of either type over the full range of loads and speeds available, and the results given here are but a selected sample to illustrate certain points relevant to the theoretical part of this paper. One of the arrangements reported here is that shown in figure 2, with just six of the M-type centre-pivoted pads in three adjacent pairs, each with a thermocouple installed. The remaining twelve pad positions were fitted with rollers below the mean pad height which enabled the self-aligning support linkage to operate normally, so that the six pads present would share the load in the manner of the design intent.

Figures 3 and 4 indicate the measured temperatures over the range of loads at two speeds 1000 rpm and 3000 rpm. The temperatures increase with load, somewhat non-linearly, but with very acceptably low values, considering that the maximum specific loading is 198 bar (2900 psi). It is noteworthy that over the full load range, the trailing pads (even numbers) experience higher temperatures than the leading pads (odd numbers), indicating the more effective cooling of the runner in the large gap between the pairs of pads than between adjacent pairs. Estimates of bearing friction loss were not obtained with any satisfactory accuracy for this small bearing since slave losses in the rig, and due to churning, greatly exceeded the bearing friction. At the two speeds considered the torque varied from zero load to full load, for the lower speed case from 6Nm to about 8.5Nm, and for the higher speed, from 10Nm to 13.5Nm indicating superficially that the bearing shear dissipation increased by about 250W (low speed) and by 1100W (high speed). On the other hand the heat carried by the circulating water, neglecting any heat conducted through the massive casing and then radiated from the rig casing, was about 80W (low speed) and about 150W (high speed). Interpreted as shear friction coefficients the above figures give ranges 0.0004-0.0013 (low speed) and 0.0002 - 0.0018 (high speed). These low friction values undoubtedly indicate the existence of hydrodynamic lubrication, even with nominally flat centre-pivoted pads. It must be therefore assumed that some convexity of the pads (of the order of some tenths of a micron) must have existed under the loaded conditions and indeed some crowning was apparent on the pads even unstressed as indicated by Talysurf. Calculation of operating film thicknesses, assuming small values of spherical crowning (of a few tenths of a micrometre) via isothermal lubrication theory and a conduction model similar to that of Ref. 2 indicate for

the maximum load of 30kN, values of $\sim 0.4\mu$ m
at the lower speed and $\sim 0.7\mu$ m at the higher
speed, and with temperatures predicted in
reasonable agreement with those measured.
Using this model the changes in power loss due
to pad shearing may be calculated to be 40W
(low speed) and 190W (high speed). For all
speeds in the range 500-4000 rpm the behaviour
was similar, with the predicted film
thicknesses at the lowest speed and highest
load being $\sim 0.3\mu$ m.

A second selection of results was with
the pad complement reduced to three thus
doubling the load/pad capability of the test
rig. Figures 5 and 6 again show the variation
of temperature with load, and show clear signs
of a developing change of operating mode.
First, however, it should be noted that for a
load of 15kN i.e. 5kN/pad, the measured
temperatures are consistently lower than for
the six pad bearing at the same pad load. (At
1000 rpm 1.1K down to 0.7K; at 3000 rpm 2.6K
down to 1.7K). At higher loads however, above
6kN/pad, temperatures rise (relatively)
dramatically, with one of the pads (No. 6)
clearly leading the way. The associated shaft
torque also showed that a distinct change in
operational mode was present, with a two to
fourfold increase in the component associated
with bearing friction. It might be noted that
reduction in load allowed the re-instatement
of the low temperature, low torque mode
immediately and subsequent examination of the
pads did not reveal significant damage, merely
minor rubbing marks at leading edge and along
the sides of the pad. No totally convincing
explanation for this behaviour can be offered
to explain the change in mode. It is evidently
not speed- or temperature-dependent. It is
however load-dependent, and it may be that the
undoubtedly significant mechanical pad
deflections which are increasingly present
distort the film shape beyond that which can
support the load hydrodynamically. There
appears at least a possibility that the
hottest pad has flipped (possibly dynamically)
to a divergent attitude from which it is
unable to recover prior to the load being
reduced. The phenomenon occurs, however, at a
specific load far in excess of that for which
a bearing design would be contemplated, and it
is therefore perhaps of limited academic
interest only.

5   CONCLUDING REMARKS

This presentation has attempted to show that:

(a) small, highly loaded ceramic bearings
     such as the Glacier-Kemel bearing may be
     analysed by reasonably standard means
     provided account is taken of heat
     transfer paths correctly.

(b) The high load capacity is obtainable in
     practice due to small, highly accurate
     components which are extremely robust.

(c) Criteria on acceptable roughness -
     operating film thickness can evidently be
     extrapolated to the sub-micron region.

6   REFERENCES

1.   ESDU Item no. 83004. 'Calculation Methods
     for Steadily Loaded, Offset Pivot,
     Tilting Pad Thrust Bearings'. March 1983.

2.   Neal, P.B. 'Heat Transfer in Pad Thrust
     Bearings' Proc. I.Mech.E. Vol 197
     p.217-228, 1982.

3.   Kamelmacher, E. 'Design and Performance
     of Silicon Carbide Product Lubricated
     Bearings'. Proc. I.Mech.E. Vol 197A
     p.257-267, 1983.

4.   Dorfman, L.A. 'Hydrodynamic Resistance
     and Heat Loss of Rotating Solids'. Oliver
     & Boyd, 1963.

ACKNOWLEDGEMENTS

The experimental work for which the results in
this paper represented a small part, was
supported by MoD(N), Glacier Rotating Plant
Bearings (Mr. A.J. Leopard and Mr. S. Gray),
GEC-ALSTHOM Engineering Systems Ltd.
(Mr. C. Rodwell) and by the then GEC
Engineering Research Centre. (Now part of
European Gas Turbines Ltd.)

The author would like to acknowledge the
contributions of his colleagues in ERC,
Dr. G.W. Beesley and Mr. K.G. Issitt during
the experimental phase.

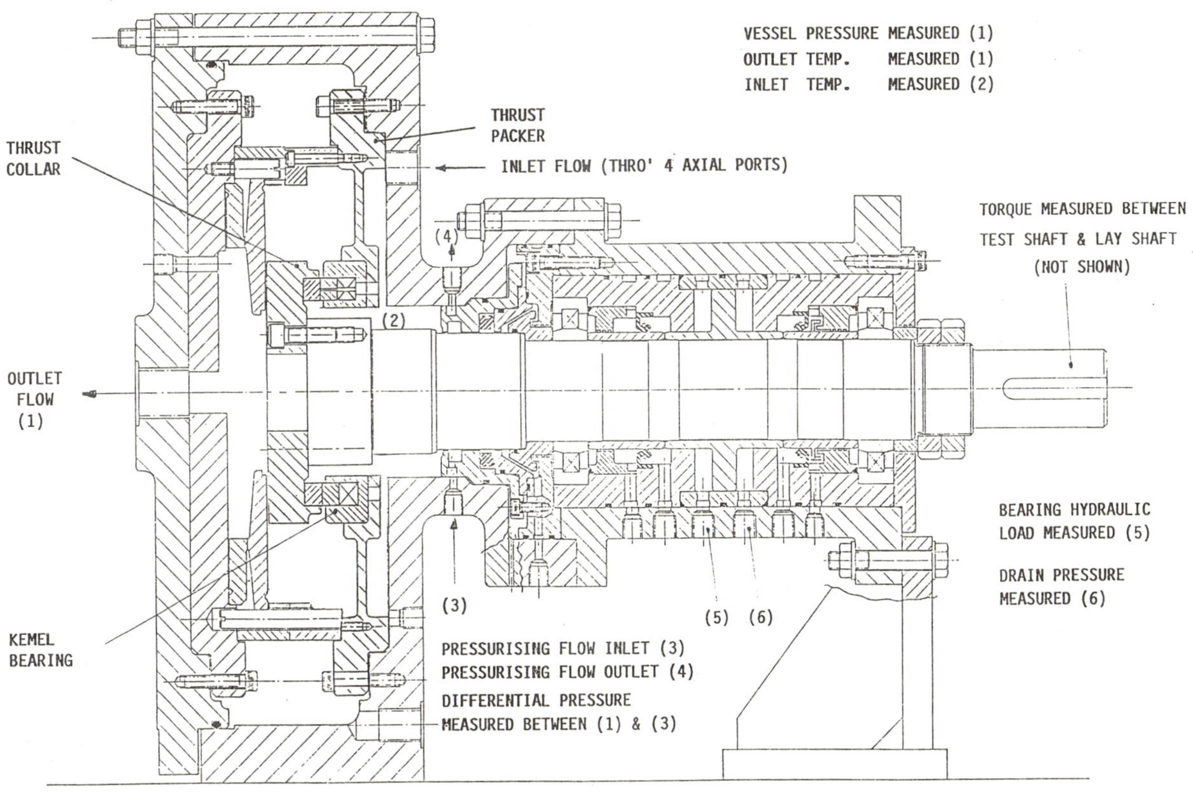

Fig 1   Cross-section of test rig

Fig 2   Test bearing

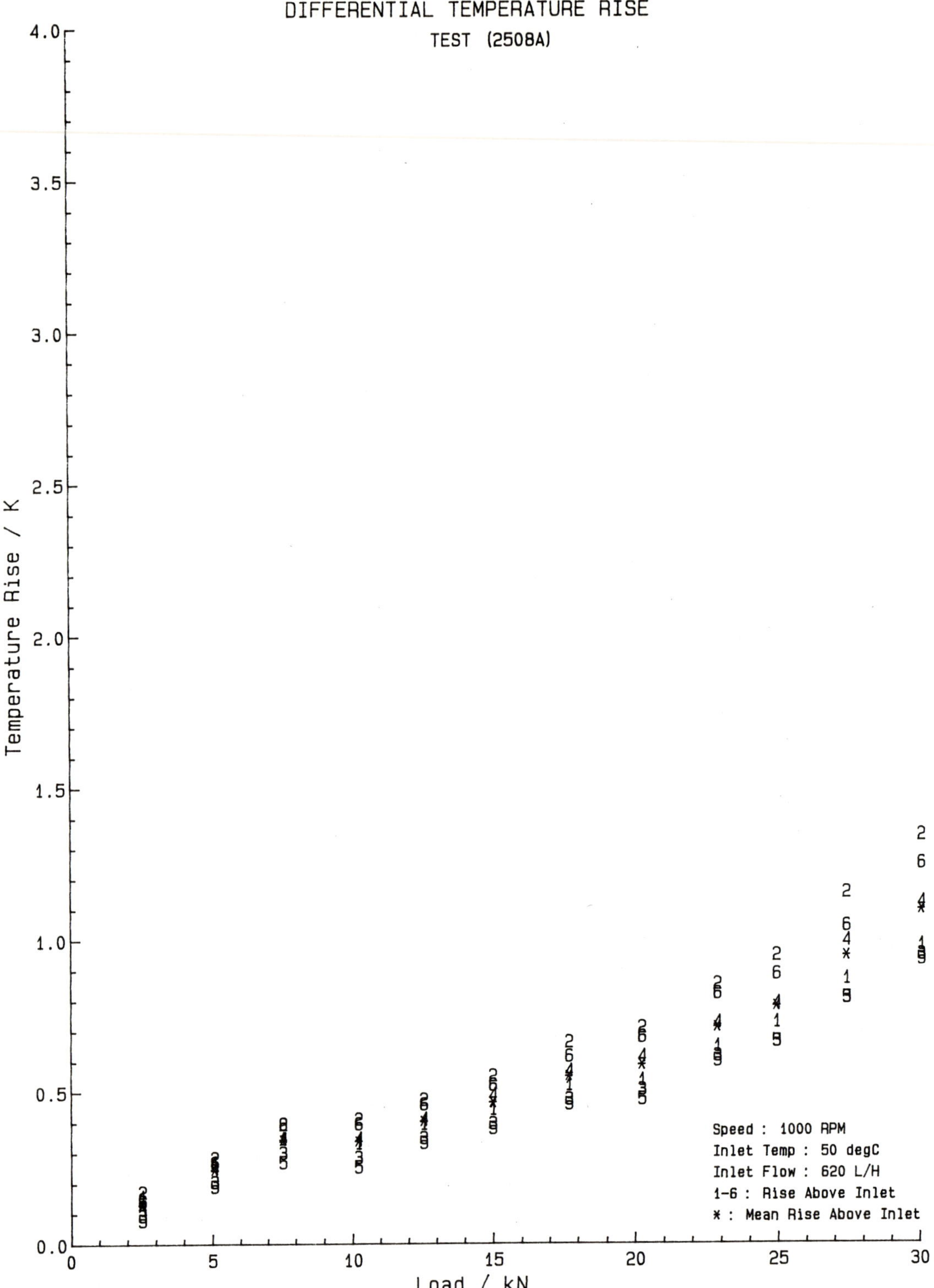

Fig 3   Temperature rise versus load
(6-pad, 1000 rpm)

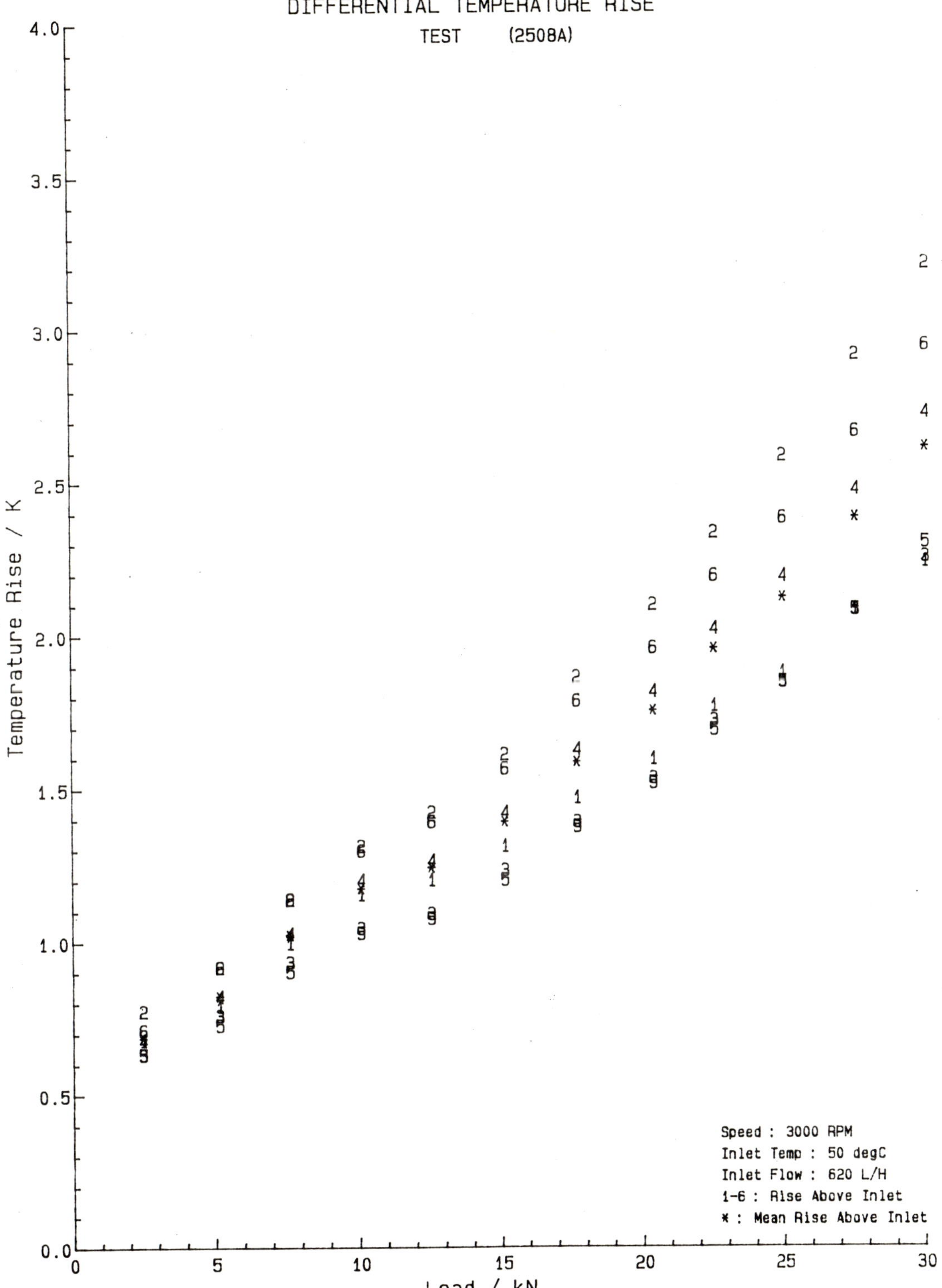

**Fig 4   Temperature rise versus load
(6-pad, 3000 rpm)**

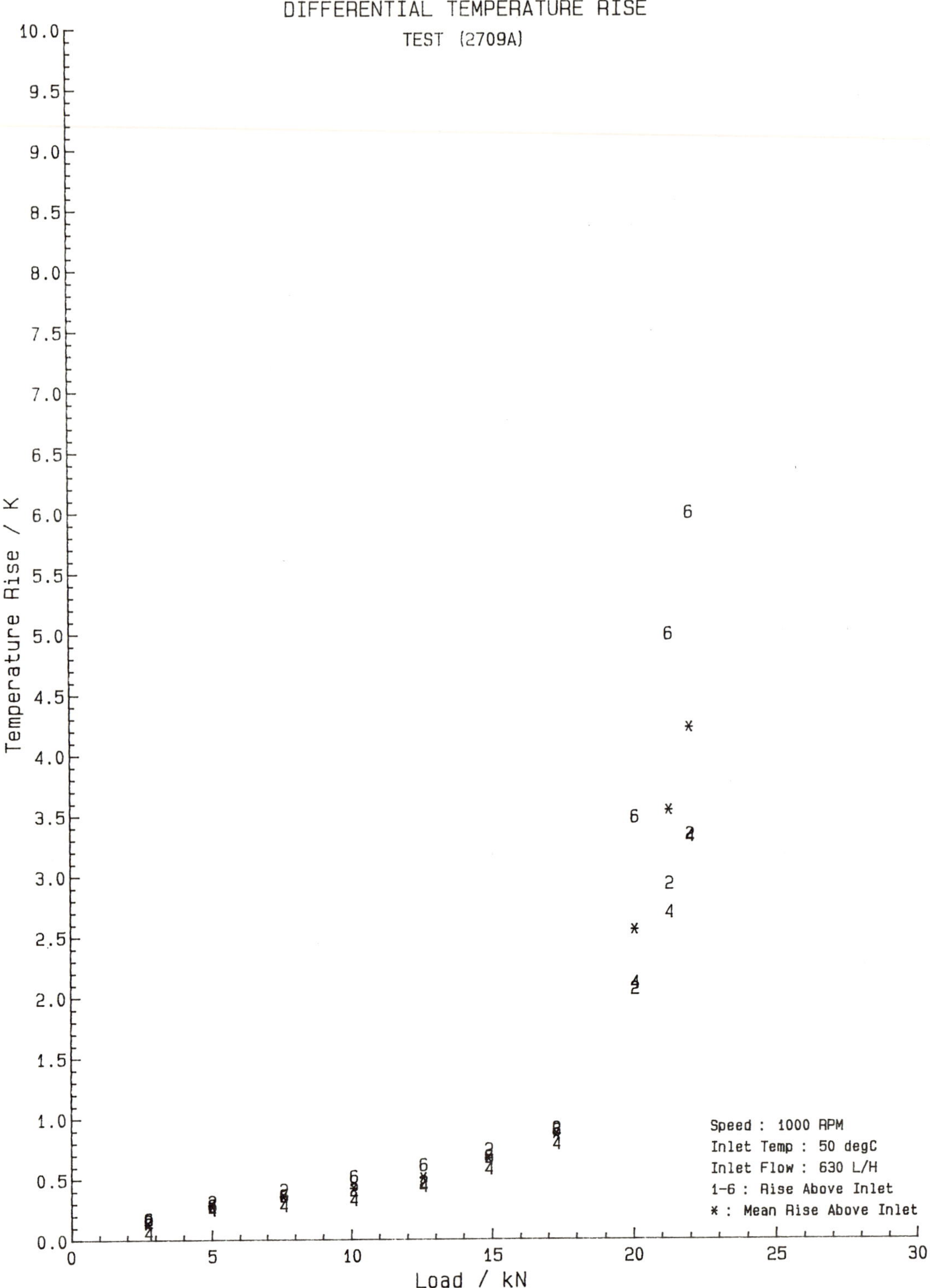

Fig 5   Temperature rise versus load
(3-pad, 1000 rpm)

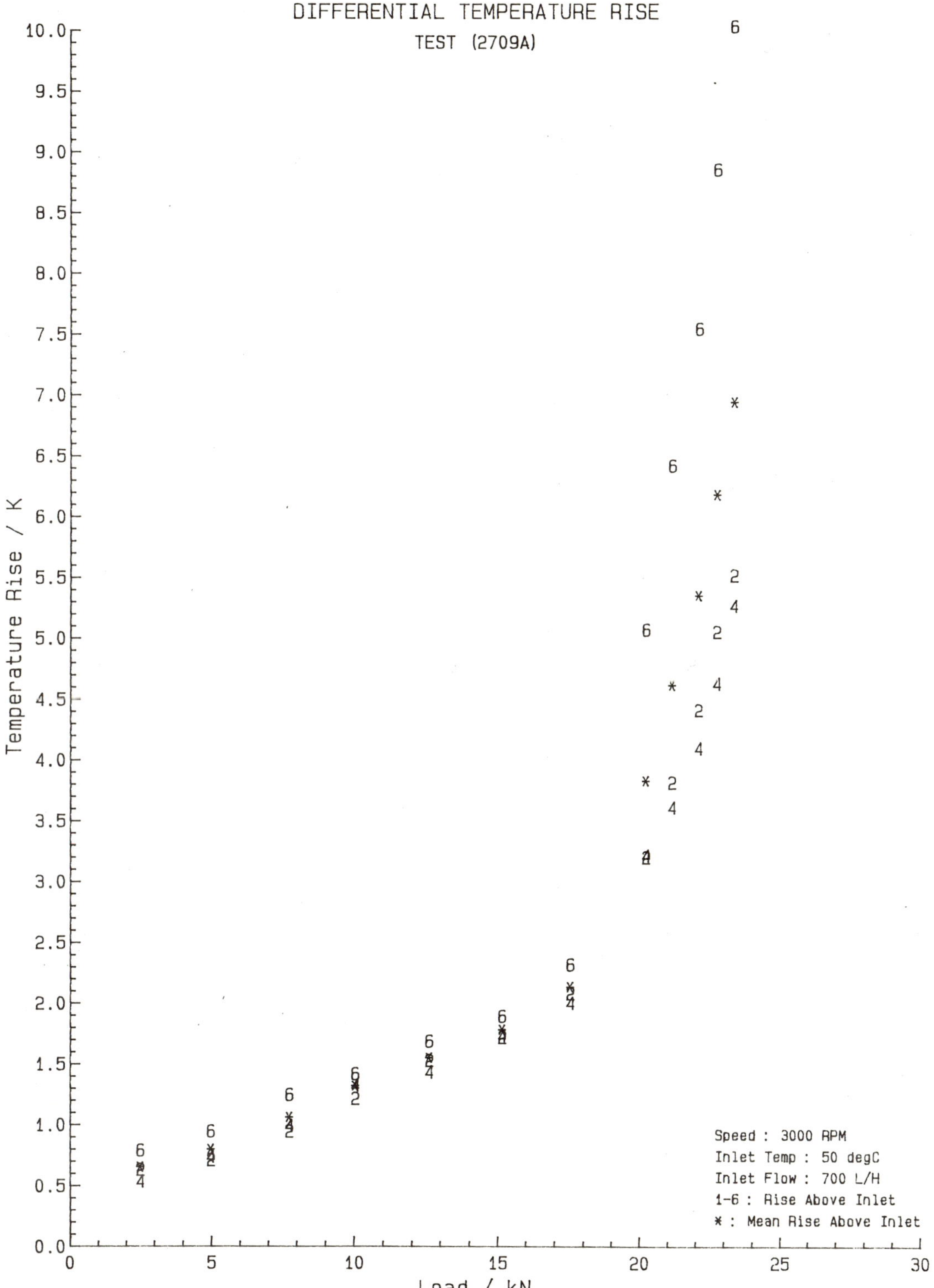

Fig 6   Temperature rise versus load
(3-pad, 3000 rpm)